Celebrating *Applied Sciences* Reaches 20,000 Articles Milestone: Invited Papers in "Applied Dentistry and Oral Sciences" Section

Celebrating *Applied Sciences* Reaches 20,000 Articles Milestone: Invited Papers in "Applied Dentistry and Oral Sciences" Section

Editor

Mary Anne Melo

MDPI • Basel • Beijing • Wuhan • Barcelona • Belgrade • Manchester • Tokyo • Cluj • Tianjin

Editor
Mary Anne Melo
General Dentistry
University of Maryland
School of Dentistry
Baltimore
United States

Editorial Office
MDPI
St. Alban-Anlage 66
4052 Basel, Switzerland

This is a reprint of articles from the Special Issue published online in the open access journal *Applied Sciences* (ISSN 2076-3417) (available at: www.mdpi.com/journal/applsci/special_issues/Milestone_Invited_Papers).

For citation purposes, cite each article independently as indicated on the article page online and as indicated below:

LastName, A.A.; LastName, B.B.; LastName, C.C. Article Title. *Journal Name* **Year**, *Volume Number*, Page Range.

ISBN 978-3-0365-7727-2 (Hbk)
ISBN 978-3-0365-7726-5 (PDF)

Cover image courtesy of Mary Anne Melo

Contents

About the Editor

Mary Anne Melo

Mary Anne Melo (DDS, MSc, Ph.D., Fellow ADM) is a Clinical Professor at the University of Maryland School of Dentistry and currently serving as Chair of the Department of General Dentistry. Dr. Melo applies her experience as a dentist and dental materials researcher to advance the development of smart and bioactive restorative materials. Her clinical areas of interest include minimally invasive dentistry, the management of high-caries-risk patients, and esthetic dentistry. Her research has focused chiefly on anticaries approaches for caries-inhibiting, antibacterial, or remineralization functionalities. Her research group has pioneered the investigations of antibacterial and remineralizing dental adhesives and resin composites. Dr. Melo is a current member of the Academy of Operative Dentistry; the International Association for Dental Research; the American Academy of Cariology, and the American Academy of Cosmetic Dentistry. Dr. Melo is a co-inventor on two patents, has edited three books, and has published more than 200 papers in the area of dental materials. Several grants support her research. She lectures nationally and internationally on diverse topics of restorative dentistry.

Preface to "Celebrating *Applied Sciences* Reaches 20,000 Articles Milestone: Invited Papers in "Applied Dentistry and Oral Sciences" Section"

We are delighted to present this special preface to celebrate a remarkable milestone in the journey of *Applied Dentistry*. With great pleasure, we announce the publication of the journal's 20,000th paper. This remarkable achievement reflects the unwavering commitment and dedication of our esteemed authors, reviewers, and readers who have contributed to the advancement of dental research.

Over the years, *Applied Dentistry* has emerged as a beacon of knowledge, serving as a platform for the dissemination of cutting-edge research and innovative ideas in various domains of dental science. Our journal encompasses a wide range of disciplines, covering everything from dental materials and biomaterials to preventive and therapeutic strategies, from clinical applications to regenerative dentistry.

To commemorate this momentous occasion, the Editors and Board Members of *Applied Dentistry* came together to extend a heartfelt invitation to researchers in all areas covered by the journal. We encouraged the submission of original research or review articles of the highest quality, showcasing the latest advancements and breakthroughs in the field of applied dentistry. This celebration is a testament to our dedication to excellence and our commitment to sharing valuable insights and discoveries with our readership.

We recognize that the strength of *Applied Dentistry* lies not only in the expertise and dedication of our Editorial Team, but also in the contributions of the vibrant research community. It is through your rigorous research, thought-provoking analyses, and innovative ideas that we have reached this significant milestone. Your dedication to advancing dental science is truly commendable, and we are honored to have played a part in bringing your research to the forefront of the field.

As we embark on this special celebration, we extend our gratitude to all the authors who have entrusted us with their valuable work, to the diligent reviewers who have provided invaluable feedback and guidance, and to our dedicated readers who have embraced the knowledge shared within the pages of *Applied Dentistry*.

Mary Anne Melo
Editor

applied sciences

MDPI

Article

In Vitro Study Comparing Retention of Custom Post and Cores Fabricated Using Conventional, CAD/CAM Milling and 3D-Printing Techniques

Hatem Alqarni [1,2,*], Abdulaziz A. AlHelal [3], Rami Jekki [4], Mathew T. Kattadiyil [4], Mohammed E. Sayed [5,6,*], Saurabh Jain [5], Seyed Aliakbar Vahdati [7,8] and Salem Dehom [9]

1 Restorative and Prosthetic Dental Science Department, College of Dentistry, King Saud Bin Abdulaziz University for Health Sciences, Riyadh 14611, Saudi Arabia
2 King Abdullah International Medical Research Center, Riyadh 14611, Saudi Arabia
3 Department of Prosthetic Dental Sciences, College of Dentistry, King Saud University, Riyadh 11545, Saudi Arabia
4 Advanced Specialty Education Program in Prosthodontics, Loma Linda University School of Dentistry, Loma Linda, CA 92350, USA
5 Department of Prosthetic Dental Sciences, College of Dentistry, Jazan University, Jazan 45142, Saudi Arabia
6 Rutgers School of Dental Medicine, Rutgers University, Newark, NJ 07103, USA
7 Advanced Specialty Education Program in Endodontics, University of California School of Dentistry, Los Angeles, CA 92350, USA
8 Private Practice, Newport Beach, CA 92660, USA
9 School of Nursing, Loma Linda University, Loma Linda, CA 92354, USA
* Correspondence: qarnih@ksau-hs.edu.sa (H.A.); drsayed203@gmail.com (M.E.S.)

Abstract: This study aimed to evaluate the differences in the retention of custom-cast non-precious post and cores (CCNPPCs) (control group), custom-milled titanium post and cores (CMTPCs), custom-printed titanium post and cores (CPTPCs), and custom-milled zirconia post and cores (CMZPCs), and to evaluate their mode of failure. The tested null hypothesis was that there were no differences in the retention of the various custom post and cores tested. A total of 80 post-and-core patterns were made using pattern resin and were divided into four groups: Group 1—fabricated via conventional casting using a non-precious casting alloy; Group 2—fabricated using a computer-aided design/computer-aided manufacturing (CAD/CAM) subtractive technique using titanium; Group 3—fabricated using a CAD/CAM additive (3D printing) technique using titanium; and Group 4—fabricated using a CAD/CAM subtractive technique using zirconia. The post and cores were cemented with resin cement and a universal pull-out test was used to check the retention. The data were statistically analyzed using one-way ANOVA tests, post hoc tests, and Tukey's adjustment for multiple comparisons. The pull-out test revealed higher retention values for CPTPCs and CMTPCs. When compared with CMZPCs, the conventional CCNPPCs revealed significantly better retention values ($p < 0.05$). Cohesive failure was observed in Groups 1, 2, and 4. However, Group 3 revealed a mixed type of failure. The CCNPPCs revealed clinically acceptable values, while the CPTPC and CMTPC groups revealed better overall values of retention and time to failure. The titanium alloy was assessed to be a promising choice for fabricating dental post-and-core restorations.

Keywords: titanium alloy; zirconia; retention; CAD/CAM; additive manufacturing; subtractive manufacturing; custom post and core; 3D printing; non-precious alloy

Citation: Alqarni, H.; AlHelal, A.A.; Jekki, R.; Kattadiyil, M.T.; Sayed, M.E.; Jain, S.; Vahdati, S.A.; Dehom, S. In Vitro Study Comparing Retention of Custom Post and Cores Fabricated Using Conventional, CAD/CAM Milling and 3D-Printing Techniques. *Appl. Sci.* **2022**, *12*, 11896. https://doi.org/10.3390/app122311896

Academic Editor: Mary Anne Melo

Received: 17 October 2022
Accepted: 19 November 2022
Published: 22 November 2022

1. Introduction

Post and cores are commonly used to restore endodontically treated teeth with the extensive coronal structural loss [1]. The retention of the post is a fundamental factor influencing definitive restoration longevity and success. Post length, shape, diameter and surface texture, and the type of cement used are associated factors that may affect the retention and stability of the post [2].

Custom-cast post and cores (CCPCs) have been reported to have superior adaptation and fit the radicular post-space walls when compared with prefabricated posts [3,4]. Compared with other post-and-core systems, CCPCs are advantageous, as they exhibit higher resistance to rotational movement forces [5], have a superior success rate [6,7], and have better retrievability for endodontic re-treatment [6,8]. Various materials can be used for fabricating custom post and cores. These include gold alloys [9–11], titanium alloy [9,11], base-metal alloys [9,12–14], zirconium oxide [11,15,16], glass fiber-reinforced composites [11,17], etc., which can be fabricated by different techniques such as casting [9–13], CAD/CAM milling [11,12,15–17], and 3D printing [11,12,14]. Various alloys are used for the fabrication of CCPCs, and most of them exhibit good retention and fracture strength [18,19]. However, CCPCs have been associated with catastrophic abutment fractures due to their high stiffness [4,5].

Custom post and cores have also been reported to be fabricated out of zirconia to overcome the esthetic limitations associated with metal CCPCs. Zirconia has been reported to have high flexural strength, high fracture toughness, chemical stability, biocompatibility, favorable optical properties, greater toughness, and maximum adaptability to the canal, as well as good esthetic characteristics [20,21]. As discussed by Baba et al. [22] and Ozkurt et al. [23], zirconia as a post-and-core material has a few disadvantages. These include more frequent root fractures (due to high rigidity) than fiber posts, decreased retention of the post (due to poor bonding between zirconia and resin cement), and poor retrievability in cases that need endodontic re-treatment.

Titanium has also gained wide acceptance in dentistry due to its biocompatibility, excellent corrosion resistance, reduced cost, ease of fabrication, and superior mechanical properties, which make it suitable as a post-and-core material [24–27]. Computer-aided design/computer-aided manufacturing (CAD/CAM) additive (3D printing) and CAD/CAM subtractive manufacturing techniques (milling) can be used to manufacture custom post and cores.

To the best of our knowledge, the current literature lacks the relevant information assessing the relationship between the retention of custom-cast non-precious post and cores (CCNPPCs), custom-milled titanium post and cores (CMTPCs), custom-printed titanium post and cores (CPTPCs) and custom-milled zirconia post and cores (CMZPCs). Therefore, this study aims to evaluate the differences in the retention of CCNPPCs (control group), CPTPCs, CMTPCs, and CMZPCs. The tested null hypothesis was that there were no differences in the retention values of the various tested custom post and cores.

2. Materials and Methods

2.1. Materials

In this study, four different techniques were used to fabricate custom post and cores (simulating the protocols for post-and-core fabrication) for the maxillary central incisors. The fabrication techniques used were: the conventional lost-wax casting technique (using a non-precious alloy), the CAD/CAM subtractive technique (using Grade 5 titanium and zirconia), and the CAD/CAM additive technique (3D printing) (using Grade 5 titanium 6AL4V). The details of each material used in the study are listed in Table 1.

2.2. Specimen Preparation

Eighty sound human maxillary central incisors, extracted for periodontal reasons, were selected for this study. The exclusion criteria included the presence of caries, restoration, root canal treatment (RCT), crack/s, attrition, very long or very short teeth, and/or severe root curve. Teeth were thoroughly cleaned with a brush after extraction, and a scalpel and a periodontal curette were used to remove any remaining hard and soft tissues from the root surfaces. Teeth were subsequently stored in 0.2% sodium azide (Merck KGaA, Frankfurter Str. 250, Darmstadt, Germany) prior to specimen preparation (ISO 28399;

2011) [28]. Each tooth was randomly assigned a number (from 1 to 80) and allocated into one of 4 groups (n = 20 each). Group 1 teeth were restored with custom-cast non-precious post and cores (CCNPPCs); Group 2 teeth were restored with custom-milled titanium post and cores (CMTPCs); Group 3 teeth were restored with custom-printed titanium post and cores (CPTPCs), and Group 4 teeth were restored with custom-milled zirconia post and cores (CMZPCs).

Table 1. Commercial names and details of materials used in the study.

Group	Material Trade Name	Manufacturer	Main Composition	Manufacturing Technique Used
Group 1	NPG	Aalba Dent, Inc., Fulton Drive, Fairfield, CA, USA	Cu, 80.7%; Al, 7.8%; Ni, 4.3%	Casting (lost-wax technique)
Group 2	KERA Ti-5 Disc	Eisenbacher Dentalwaren ED GmbH Dr.-Konrad-Wiegand-Straße, Wörth am Main, Germany	Ti, 89%; Al, 6.4%; V, 4.1%	Milling
Group 3	Ti-6Al-4V	Renovis Surgical, West Lugonia Ave, Austin, TX, USA	Ti, 89%; Al, 6.4%; V, 4.1%	3D printing
Group 4	BruxZir Full-Strength Zirconia	BruxZir; Glidewell Laboratory Inc., Newport Beach, CA, USA	Monolithic zirconia (zirconium oxide)	Milling

2.2.1. Mounting Teeth in Acrylic-Resin Blocks

Teeth were individually mounted in a special specimen holder using epoxy resin (Exakto-form; Bredent, Derbyshire, UK) (with an elastic modulus of 12 GPa, which is similar to the elastic modulus of human bone (18 GPa)) [29] with the use of a test mount former of 2 cm^3 in dimension 2–3 mm below the cementoenamel junction (CEJ). A prefabricated jig was used to position each tooth in the test mount former during the immersion of the tooth in acrylic resin to standardize the tooth position to be centralized within the test mounts. The test mount jig was used to standardize the tooth position while performing tooth preparation and RCT (Figure 1). To avoid the dehydration of natural teeth due to the heat generated during the polymerization of acrylic resin, the resin block was cooled in water [30]. The teeth were prepared to have a 2 mm ferrule and a 1 mm shoulder finish line (Figure 2) [31–33].

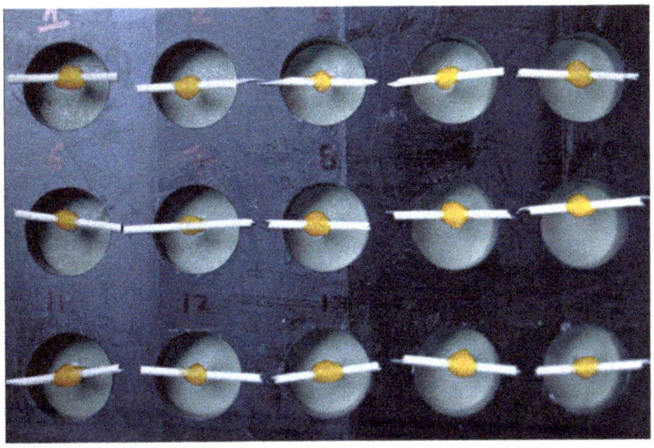

Figure 1. Teeth were mounted in acrylic-resin blocks using a test mount former.

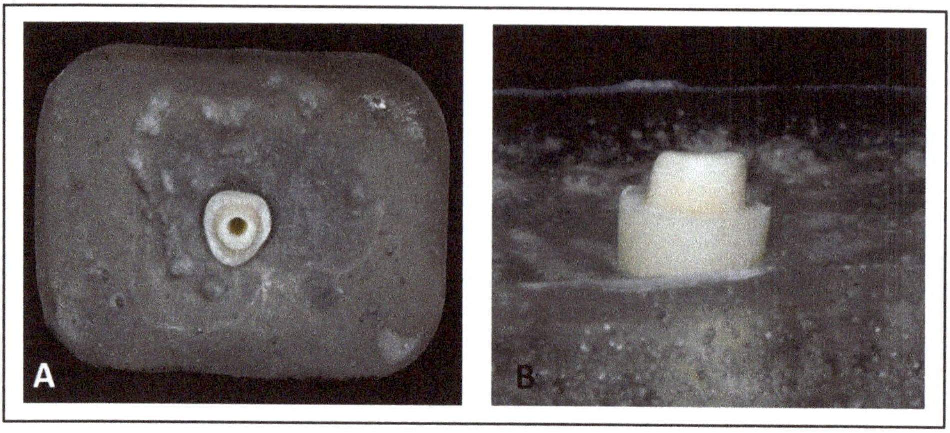

Figure 2. Preparation of tooth to have a 2 mm ferrule and a 1 mm shoulder finish line. (**A**) Occlusal view. (**B**) Lateral view.

2.2.2. Root-Canal Preparation and Obturation

Access-cavity preparation was performed using a size 2 diamond round bur on a high-speed handpiece with copious water. The working length was established as 1 mm shorter than the root apex. Root-canal preparation for each tooth was performed up to size 40/0.06 with a Vortex Blue rotary file (Dentsply Sirona, Tulsa, OK, USA) in a crown-down fashion. Canals were irrigated with NaOCl 5.25% and EDTA 17% (chelating agent) (Ultradent Products Inc., South Jordan, UT, USA) to remove organic and inorganic debris and smear layers [34]. Upon completion of the cleaning and shaping and before obturation, the prepared root canals were dried with sterile paper points (Sure-endo, Gyeonngi-do, Republic of Korea). A matching size 40/0.06 gutta-percha master cone (Dentsply Sirona, Tulsa, OK, USA) coated with AH Plus sealer (Dentsply Sirona GmbH, De-Trey-Straße, Konstanz, Germany) was used for obturation [34]. Root canals were obturated using the warm vertical compaction technique with System-B and Obtura (Kerr/Sybron Endo Corp., Brea, CA, USA).

2.2.3. Post-Space Preparation

To obtain a standardized length for the posts, the coronal portion of gutta-percha was removed with System-B (Kerr corporation, Brea, CA, USA) until an adequate length (11 mm) was achieved [35–38]. Definitive post length and width were prepared and established with the use of Peeso reamers (Maillefer S.A., Ballaigues, Switzerland) up to size 3. Each canal was then cleaned using an air/water spray and EDTA to remove debris and then dried with paper points.

2.2.4. Post-and-Core Fabrication

Eighty custom post-and-core patterns were fabricated with the use of auto-polymerizing acrylic resin (Pattern Resin LS; GC America, Alsip, IL, USA) and serrated plastic posts. Each serrated plastic post was relined with acrylic resin and then inserted into the root canal until the canal space was bound to the walls of the prepared teeth. Then, the core was built up using a GC pattern and prepared using a diamond bur (ISO No. 010; Brasseler USA Dental, Georgia) to achieve a core with a 4 mm height (Figure 3). A hole was made in the coronal part of the GC pattern core and was used to attach the post and core to an Instron machine (Instron; Model 5585H; Instron Corp., Norwood, MA, USA) during the pull-out test.

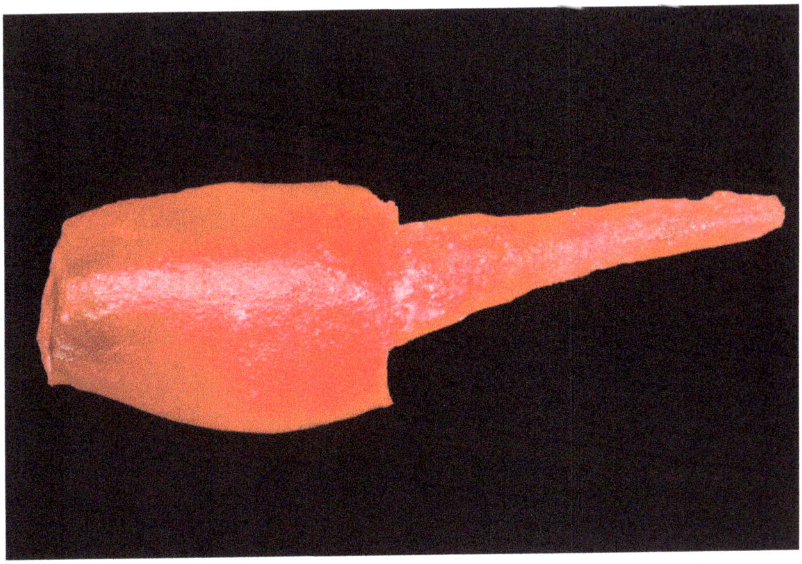

Figure 3. Pattern resin post-and-core build-up.

Specimens with a post-and-core pattern were randomly divided into 4 groups. Twenty acrylic-resin pattern posts were cast with a non-precious alloy using the conventional lost-wax casting technique. Sixty acrylic-resin pattern posts were scanned with a desktop scanner (3Shape D900L; 3Shape Dental System, Copenhagen, Denmark). The STL files for each scanned post and core were sent to Core 3D and Renovis for the fabrication of 20 milled titanium (Grade 5) (Ti6AL4V), 20 printed titanium (Grade 5) (Ti6AL4V), and 20 milled zirconia post and cores (Figure 4). The prepared specimens were stored with 100% humidity at room temperature to simulate the in vivo humidity until they were returned for testing [39] (Figure 5).

Figure 4. Sample of STL file of the proposed design of the specimen.

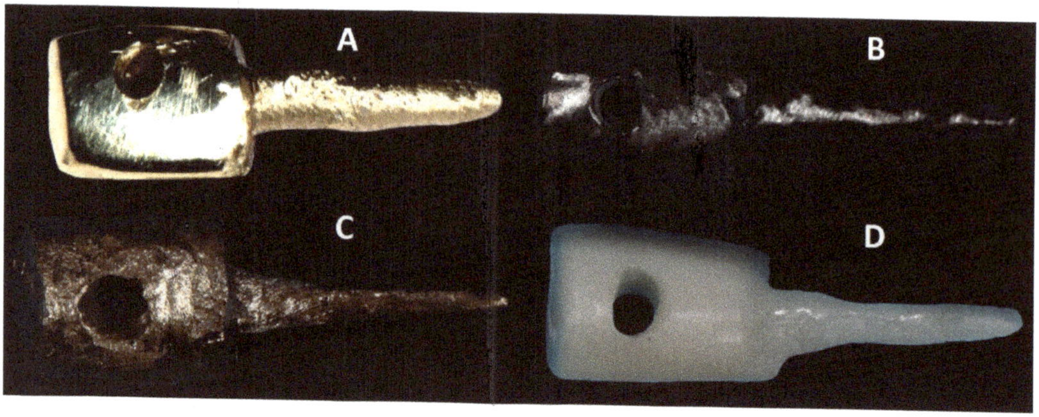

Figure 5. (**A**) Custom-cast non-precious post and core. (**B**) Custom-milled titanium post and core. (**C**) Custom-printed titanium post and core. (**D**) Custom-milled zirconia post and core.

2.2.5. Surface Treatment of Post and Core

According to the manufacturer's recommendation, Group 1 (CCNPPCs), Group 2 (CMTPCs), and Group 3 (CPTPCs) post and cores were first treated using ultrasonic cleaning solution in 96% isopropyl (3 min) and were then airborne-particle abraded (50 μm Al_2O_3 at 2.8 bar for 5 s). Group 4 (CMZPCs) post and cores were treated using ultrasonic cleaning in 96% isopropyl (3 min) and Rocatec soft (30 μm airborne-particle abrasion at 2.8 bar for 12 s) over the entire zirconia surface; finally, silane coupling agent (Espe-Sil; 3M ESPE; Seefeld, Germany) was applied [40].

2.2.6. Post-and-Core Cementation

Resin luting cement, Rely X Unicem resin cement (3M ESPE; Seefeld, Germany), was used to cement all post and cores. The cement was mixed according to the manufacturer's instructions and was coated on the post. Cement was also applied into the root canals by attaching an elongation tip to the nozzle. After that, posts were gently inserted into the root canals to reduce the hydrostatic pressure; they were positioned in place under firm finger pressure, and excess cement was removed (Figure 6). Then, specimens of each group (n = 20) were kept in normal saline for 24 h in a refrigerator before testing. Thermocycling (Thermocycling test apparatus; Durant) was performed for all specimens in 2 water tanks (cold, warm) by immersion at the temperatures of 5 °C and 55 °C and fixed time intervals (16 s cold, 16 s warm) for a total of 6000 cycles, which represented 7 months of clinical use [41].

2.3. *Placing Specimens on the Measuring Machine (Instron Testing Machine)*

Each tooth with a post and core was subjected to a pull-out test by a universal Instron testing machine (Instron Corp.) at a crosshead speed of 0.5 mm/min. The device was calibrated before placing each sample.

To standardize the location and direction of the specimens in the machine, samples were placed in a custom-made, self-aligning device. A hook-shaped attachment was passed through the hole created in the custom post and core, which was attached to the Instron testing machine (ITM). The acrylic-resin blocks held the teeth securely during retention testing (Figure 7). Force was applied until the cemented post was removed from the prepared post space. The retention values were recorded as the amount of force required to dislodge the post and core from the prepared post space [30]. Test specimens were considered to have failed when the post and core separated from the tooth.

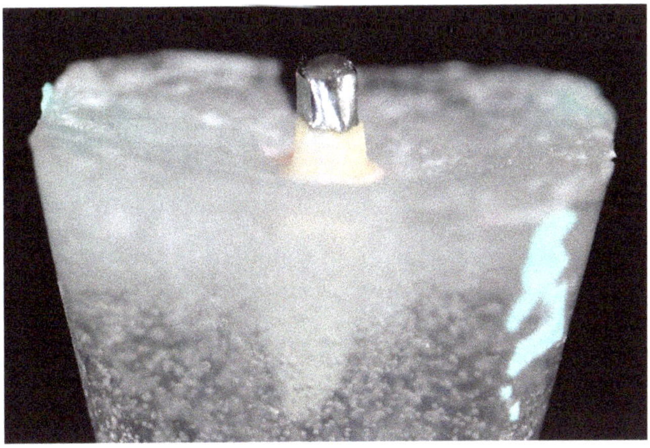

Figure 6. View of a cemented post and core.

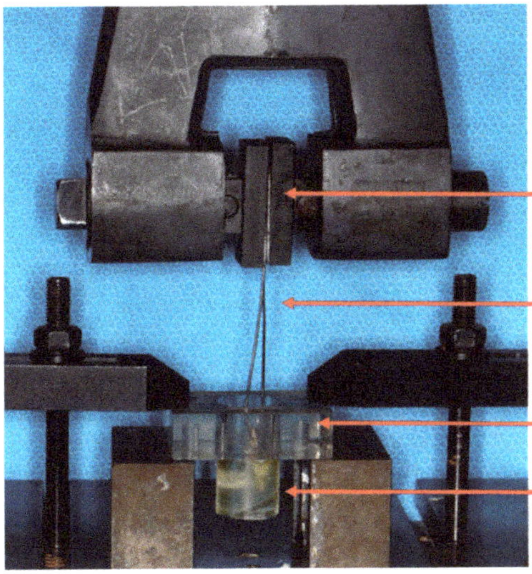

Figure 7. Specimen of mounted post and core in testing assembly for retention test.

2.4. Scanning Electron Microscopy (SEM) Analysis

The modes of failure were classified as follows: (1) adhesive (clean break at the bond), (2) cohesive (full break in the base material or tooth), and (3) mixed (combination of adhesive and cohesive failure modes). A precision low-speed Techcut 4tm saw (ALLIED High-Tech Product, Inc., Compton, CA, USA) was used to slice the roots along the long axis under constant water irrigation. A stereomicroscope at 50× magnification and a scanning electron microscope (SEM) (FEI QUANTA 250 FEG, Thermo Fisher Scientific, Waltham, MA, USA) were used to examine the sectioned root canals, and displaced posts were examined using a stereomicroscope at the same magnification [42].

2.5. Statistical Analysis

The descriptive statistics were reported as means ± standard deviation and medians with minimum and maximum values for all variables for each dental-material group. The one-way ANOVA procedure was used to test if there was a difference in total time and average load among the treatment groups [43]. Post hoc tests were performed using Tukey's adjustment for multiple comparisons. The Games–Howell test was used for multiple comparisons instead of Tukey's test if the assumption of equal variances was not assumed.

The independent-sample Mann–Whitney U test with Bonferroni adjustment was used to compare the medians of the displacement scores of each group with the gold-standard control group. Statistical analyses were performed using IBM SPSS Statistics (Version 25; IBM Corporation 1989, 2018. IBM Corp., Armonk, NY, USA).

3. Results

Descriptive statistics are given as means with 95% confidence intervals (CIs) for the variables of maximum total time, average total time, average load, and maximum load for each group (Table 2).

Table 2. Means with 95% confidence intervals for each factor and differences in clinical factors by the group.

Characteristic	Group 1	Group 2	Group 3	Group 4	*p*-Value	Differences in Clinical Factors by the Group (Mean Difference) (95% CI)
	Mean (95% CI)	Mean (95% CI)	Mean (95% CI)	Mean (95% CI)		
Total Time—Average (Seconds)	130.3 (112.6, 148.1)	180.4 (147.7, 213.0)	274.0 (217.3, 330.6)	109.6 (84.0, 135.2)	<0.001	Total time Average * Group 1 vs. Group 2: 20.8 (−19.4, 61.0) Group 1 vs. Group 3: −143.6 (−222.2, −65.1) # Group 1 vs. Group 4: −50.0 (−98.3, 1.7) # Group 2 vs. Group 3: −164.4 (−245.8, −83.1) # Group 2 vs. Group 4: −70.8 (−124.2, −17.4) # Group 3 vs. Group 4: 93.6 (8.8, 178.5) #
Total Time—Max. (Seconds)	261.6 (226.1, 297.1)	360.7 (295.4, 426.0)	547.9 (434.6, 661.1)	219.2 (168.0, 270.3)	<0.001	Total Time Max. * Group 1 vs. Group 2: 42.5 (−37.9, 122.8) Group 1 vs. Group 3: −286.2 (−443.3, −129.2) # Group 1 vs. Group 4: −99.1 (−195.8, −2.4) # Group 2 vs. Group 3: −328.7 (−491.4, −166.0) # Group 2 vs. Group 4: −141.6 (−248.3, −34.8) # Group 3 vs. Group 4: 187.1 (17.4, 356.8) #
Average Load (N)	131.1 (118.2, 144.0)	150.9 (133.5, 168.2)	156.5 (142.8, 170.1)	96.9 (83.4, 110.5)	<0.001	Average Load ** Group 1 vs. Group 2: −5.6 (−31.2, 20.1) Group 1 vs. Group 3: 19.7 (−5.9, 45.4) Group 1 vs. Group 4: 53.9 (28.3, 79.6) Group 2 vs. Group 3: 25.3 (−0.30, 51.0) Group 2 vs. Group 4: 59.5 (33.9, 85.2) # Group 3 vs. Group 4: 34.2 (8.5, 59.8) #
Max. Load (N)	295.9 (261.0, 330.8)	302.7 (273.1, 332.3)	361.5 (309.8, 413.2)	248.1 (202.9, 293.3)	<0.002	Max. Load ** Group 1 vs. Group 2: −6.8 (−65.6, 52.0) Group 1 vs. Group 3: −65.6 (−146.2, 15.0) Group 1 vs. Group 4: 47.8 (−25.8, 121.3) Group 2 vs. Group 3: −58.8 (−132.0, 14.5) Group 2 vs. Group 4: 54.6 (−18.6, 127.8) Group 3 vs. Group 4: 113.4 (40.1, 132.0) #

Max.: maximum; CI, confidence interval; Group 1 (CCNPPCs), Group 2 (CMTPCs), Group 3 (CPTPCs), Group 4 (CMZPCs); * Games–Howell test; ** Tukey's adjusted post hoc test; # *p*-value significant at $p < 0.05$.

Using the one-way ANOVA with Tukey's adjustment, significant differences in the average-load variables were observed between the four different groups, with higher values recorded for Group 3 (CPTPC) in comparison with all other groups (Figure 8).

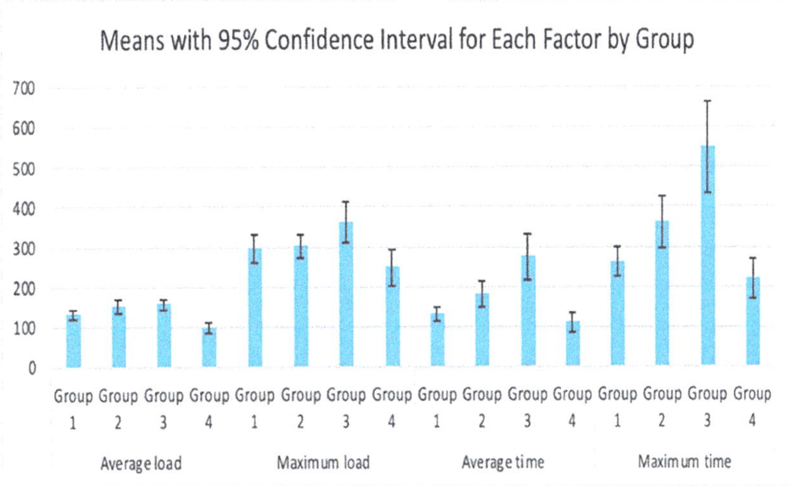

Figure 8. Comparison between fabrication materials with the control group.

The mean difference values are shown in Table 2. The load values necessary to create failure in Group 4 (CMZPCs) were lower when compared with those of Group 3 (CPTPCs), Group 2 (CMTPCs), and Group 1 (CCNPPCs); furthermore, all groups showed different distributions at the time of failure and were statistically significantly higher for Group 3 (CPTPCs) and Group 2 (CMTPCs) ($p < 0.05$). Group 3 (CPTPCs) and Group 1 (CCNPPC group) required loads of 361.5 N and 295.9 N, respectively, to fail, and the difference was statistically significant ($p < 0.05$). When considering Group 4 (CMZPC group) and Group 3 (CPTPCs), they required loads of 361.5 N and 248.1 N, respectively, and this difference was statistically significant ($p < 0.05$). The differences in the averages of maximum time and average total time between Group 3 (CPTPCs), Group 2 (CMTPCs), Group 1 (CC-NPPCs), and Group 4 (CMZPCs) were significant ($p < 0.05$), with higher values recorded for Group 3 (CPTPCs) in comparison with all other groups (Figure 8).

Using the Mann–Whitney U test with Bonferroni adjustment, no significant differences in the medians of the displacement scores were observed between each group and the control group (Group 1), as shown in Table 3. All Bonferroni adjusted p-values were >0.05.

Table 3. Comparison of different post-and-core fabrication materials with the control group.

	Group 1	**Group 2**	**Group 3**	**Group 4**
	Median (min–max)	Median (min–max)	Median (min–max)	Median (min–max)
Displacement (mm)	0.2 (0.1–1.0)	0.5 (0.1–1.0)	0.2 (0.1–0.8)	0.1 (0.1–1.0)

p-value significant at $p < 0.05$.

The failure mode varied between the groups, according to the scanning-electron-microscopy images. The majority of failures for all tested custom post-and-core systems were cohesive failures, except for Group 4, which revealed a mixed type of failure, as shown in Figures 9–12.

Figure 9. Scanning-electron-microscope image of (**A**) Sectioned root canal, and (**B**) displaced post, at 200× magnification, showing cohesive failure mode in Group 1 (CCNPPCs) sample at root/post interface.

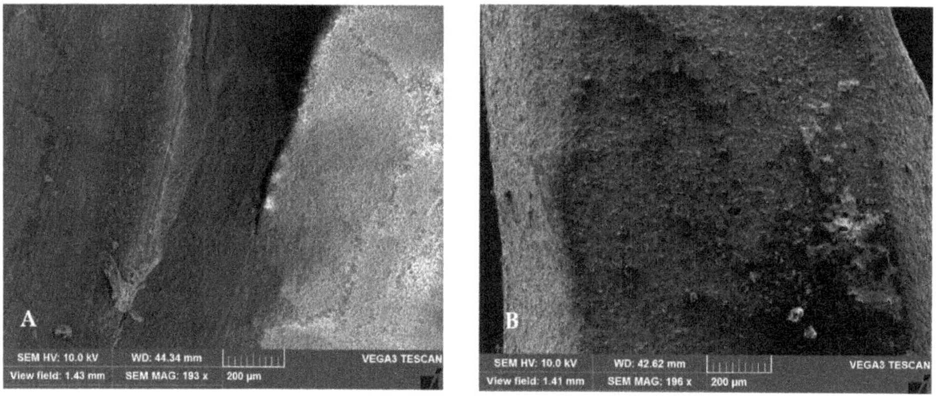

Figure 10. Scanning-electron-microscope image of (**A**) Sectioned root canal, and (**B**) displaced post, at 200× magnification, showing cohesive failure mode in Group 2 (CMTPCs) sample at root/post interface.

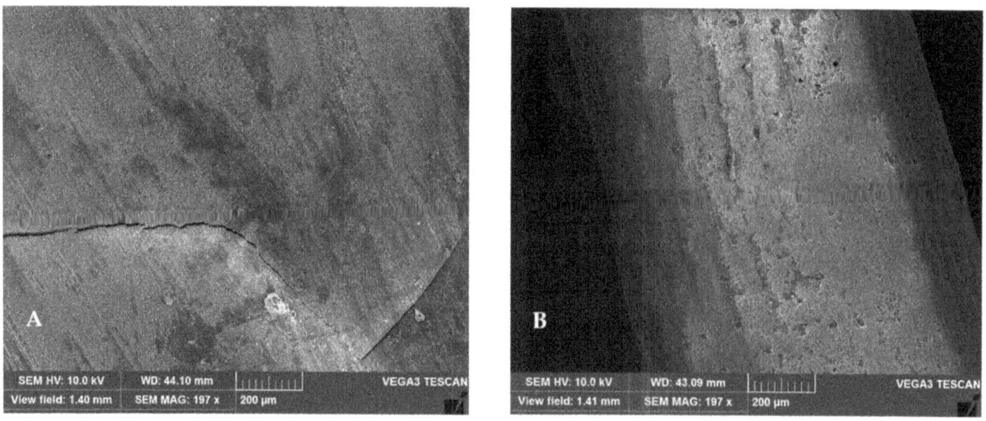

Figure 11. Scanning-electron-microscope image of (**A**) Sectioned root canal, and (**B**) displaced post, at 200× magnification, showing cohesive failure mode in Group 3 (CPTPCs) sample at root/post interface.

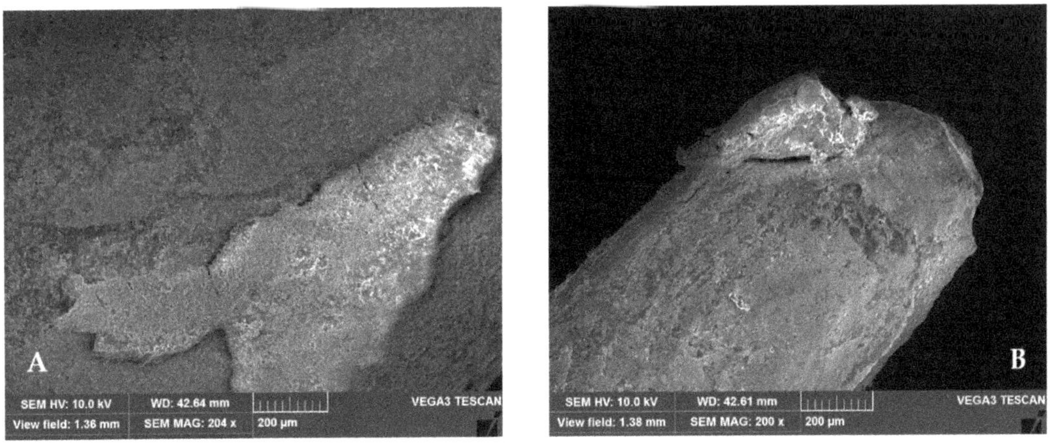

Figure 12. Scanning-electron-microscope image of (**A**) Sectioned root canal, and (**B**) displaced post, at 200× magnification, showing cohesive failure mode in Group 4 (CMZPCs) sample at root/post interface.

4. Discussion

This study evaluated the differences in the retention of different types of custom post and cores. Significant differences in retention were observed when the groups were compared. Therefore, the tested null hypothesis could be rejected; however, retention varied according to the fabrication technique and material used for manufacturing custom post and cores.

Several studies have reported acceptable clinical outcomes with CCNPPCs, which have been used for many years; hence, they were used as the control group in this study [2,4–8,18]. Advancements in materials science and the use of CAD/CAM have transformed the approach to fabricating indirect restorations in the field of prosthodontics, including the fabrication of custom post and cores. These new materials are easier to fabricate when compared with non-precious alloys. The current literature lacks evidence to support that the newly introduced custom-printed and -milled titanium post and cores offer comparable retention and effectiveness.

Various studies have reported that the success of teeth restored with custom post and cores correlated with the design, technique, and material used for the fabrication of the post and cores [6,7,42,44,45]. Recently, Wei Liu et al. found that post and cores fabricated with the CAD-CAM milled technique using a cobalt–chromium alloy could be an alternative to conventional casting for metal-post-and-core fabrication [46]. However, the retention of CAD/CAM milled or printed post and cores was neither evaluated nor tested.

The results of the present study showed that there was a significant difference ($p < 0.05$) between the retention of teeth restored with custom-printed and -milled titanium post and cores and that of teeth restored with CCNPPCs and CMZPCs. The retention difference between the two groups of custom-printed and -milled titanium post and cores was not significant ($p > 0.05$), and the CMZPC group revealed significantly lower retention ($p < 0.05$). The possible reason for the difference between custom post and cores and the retention of the zirconia group could lie in the post-surface configuration and roughness of the printed titanium post, which allows the post-and-core material to form micromechanical retention locks, whereas the smooth surface of zirconia reduces mechanical retention. Maya et al. reported similar results, with metal posts revealing significantly greater retention (495.5 N $\pm$ 75.9 N) than zirconia posts (241 $\pm$ 89.3 N) [47]. In addition, the study by Cohen et al. reported that zirconia posts had extremely low retention values (104.5 $\pm$ 34.8) [48]. These findings are consistent with the results of our study. Previous research has already demonstrated that resin cement can provide greater retention than non-resin cement [49,50]. However, this observation was not confirmed in this study, as we did not evaluate post and cores cemented with non-resin cements.

The finishing and polishing of post-and-core surfaces may improve the fit; however, this could affect the retention. It was noted that among the custom post and cores, those that were made from printed titanium required less adjustment and showed a better fit than the other groups. Studies have reported that differences in retention may also be due to differences in the adaptation of the posts [51–53]. Poor adaptation can lead to an increase in resin-matrix cement layer thickness which can increase in cracks, pores, and micro-spaces. These structural defects can cause stress concentration leading to a reduction in interface strength [51–53]. Amin et al. [54] reported that a higher volume of resin cement can also cause higher polymerization shrinkage leading to poor bond strength.

To date, due to the complexity of casting post and cores and the wide range of available materials, few studies have quantitatively evaluated the fit and accuracy of custom post and cores using micro-CT scanning, nor have they evaluated the fit and accuracy of custom post and cores via a visual inspection or direct measurements of the gap filled with cement material, as the latter has only been used to evaluate the internal fit and adaption of dental restorations [55,56].

The limitations of this study and recommendations for future ones are listed below:

1. Thermocycling was used for a short period. Further studies with longer thermocycling periods should be conducted;
2. In the current study, only one type of luting cement was used. Additional studies using different types of cements should be performed to assess the effect of the type of cement on the retention of these post and cores;
3. The effects of saliva and temperature changes in the oral cavity were not replicated in this study. The simulated clinical situations might have affected the results; hence, further studies that simulate the oral environment are recommended.

5. Conclusions

Within the limitations of the present study and based on the findings, it can be concluded that custom-printed and -milled titanium post and cores revealed significantly higher retention and bonding to resin cement than CMZPCs and similar to the control group. The CMZPCs revealed the lowest retention and adhesion to the resin cement. Therefore, the use of titanium in the fabrication of printed and milled post and cores was effective in terms of retention and could be an alternative material of choice to non-precious alloys. Further studies with in vivo testing are required to confirm the conclusions of this study.

Author Contributions: Conceptualization, H.A., R.J., M.T.K., S.A.V. and S.D.; methodology, H.A., R.J., M.T.K., S.A.V., S.D. and A.A.A.; software, H.A., S.A.V., S.D. and A.A.A.; validation, H.A., R.J., M.T.K., S.A.V. and S.D.; formal analysis, H.A., R.J., M.T.K., S.A.V., S.D. and A.A.A.; investigation, H.A., R.J., M.T.K., S.A.V. and S.D.; resources, H.A., R.J., M.T.K., S.A.V., M.E.S., S.J. and S.D.; data curation, H.A., R.J., M.T.K., S.A.V. and S.D.; writing—original draft preparation, H.A., R.J., M.T.K., S.A.V., S.D., M.E.S., S.J. and A.A.A.; writing—review and editing, H.A., R.J., M.T.K., S.A.V., S.D., M.E.S., S.J. and A.A.A.; visualization, H.A., R.J., M.T.K., S.A.V., M.E.S., S.J. and S.D., supervision, H.A., R.J., M.T.K., S.A.V. and S.D.; project administration, H.A., S.A.V., S.D., M.E.S., S.J. and A.A.A. All authors have read and agreed to the published version of the manuscript.

Funding: This research received no external funding.

Institutional Review Board Statement: IRB# 5220414.

Informed Consent Statement: Not applicable.

Data Availability Statement: The data that support the findings of this study are available from the corresponding authors upon reasonable request.

Acknowledgments: The authors thank and acknowledge Ozgul Yasar-Inceoglu (California State University, Chico, CA, USA) for support with scanning electron microscopy in this investigation.

Conflicts of Interest: The authors declare no conflict of interest.

References

1. The Glossary of Prosthodontic Terms: Ninth Edition. *J. Prosthet. Dent.* **2017**, *117*, e1–e105. [CrossRef]
2. Raedel, M.; Fiedler, C.; Jacoby, S.; Boeing, K.W. Survival of teeth treated with cast post and cores: A retrospective analysis over an observation period of up to 19.5 years. *J. Prosthet. Dent.* **2015**, *114*, 40–45. [CrossRef] [PubMed]
3. Goto, Y.; Nicholls, J.I.; Phillips, K.M.; Junge, T. Fatigue resistance of endodontically treated teeth restored with three dowel-and-core systems. *J. Prosthet. Dent.* **2005**, *93*, 45–50. [CrossRef]
4. Schillingburg, H.T.; Hobo, S.; Whitsett, L.D.; Jacobi, R.; Brackett, S.E. *Fundamentals of Fixed Prosthodontics*, 3rd ed.; Quintessence Publishing: Chicago, IL, USA, 1997; p. 40.
5. Fernandes, A.S.; Shetty, S.; Coutinho, I. Factors determining post selection: A literature review. *J. Prosthet. Dent.* **2003**, *90*, 556–562. [CrossRef]
6. Bergman, B.; Lundquist, P.; Sjögren, U. Restorative and endodontic results after treatment with cast posts and cores. *J. Prosthet. Dent.* **1989**, *61*, 10–15. [CrossRef]
7. Creugers, N.H.; Mentink, A.G.; Käyser, A.F. An analysis of durability data on post and core restorations. *J. Dent.* **1993**, *21*, 281–284. [CrossRef]
8. Pontius, O.; Hutter, J.W. Survival rate and fracture strength of incisors restored with different post and core systems and endodontically treated incisors without coronoradicular reinforcement. *J. Endod.* **2002**, *28*, 710–715. [CrossRef]
9. Venkataraman, K.J.; Thanapathi, S.; Balasubramanian, S.; Gandhi, S.A.; Sarojinikutty, A.C. Fracture Resistance of Titanium, Chrome-Cobalt, and Gold Alloy as Post and Core Materials: A Comparative Evaluation. *J. Pharm. Bioallied. Sci.* **2020**, *12* (Suppl. S1), S583–S588. [CrossRef] [PubMed]
10. Jung, R.E.; Kalkstein, O.; Sailer, I.; Roos, M.; Hämmerle, C.H. A comparison of composite post buildups and cast gold post-and-core buildups for the restoration of nonvital teeth after 5 to 10 years. *Int. J. Prosthodont.* **2007**, *20*, 63–69.
11. Al-Qarni, F.D. Customized Post and Cores Fabricated with CAD/CAM Technology: A Literature Review. *Int. J. Gen. Med.* **2022**, *15*, 4771–4779. [CrossRef] [PubMed]
12. Piangsuk, T.; Dawson, D.V.; El-Kerdani, T.; Lindquist, T.J. The Accuracy of Post and Core Fabricated with Digital Technology. *J. Prosthodont.* **2022**, *ahead of print*. [CrossRef] [PubMed]
13. Khiavi, H.A.; Habibzadeh, S.; Safaeian, S.; Eftekhar, M. Fracture Strength of Endodontically treated Maxillary Central Incisors restored with Nickel Chromium and Nonprecious Gold Alloy Casting Post and Cores. *J. Contemp. Dent. Pract.* **2018**, *19*, 560–567. [CrossRef] [PubMed]
14. Kanduti, D.; Korat, L.; Kosec, T.; Legat, A.; Ovsenik, M.; Kopač, I. Comparison between accuracy of posts fabricated using a digital CAD/CAM technique and a conventional direct technique. *Int. J. Prosthodont.* **2021**, *34*, 212–220. [CrossRef] [PubMed]
15. Libonati, A.; Di Taranto, V.; Gallusi, G.; Montemurro, E.; Campanella, V. CAD/CAM customized glass fiber post and core with digital intraoral impression: A case report. *Clin. Cosmet. Investig. Dent.* **2020**, *12*, 17–24. [CrossRef] [PubMed]
16. Lee, J.H. Fabricating a custom zirconia post-and-core without a post-and-core pattern or a scan post. *J. Prosthet. Dent.* **2018**, *120*, 186–189. [CrossRef] [PubMed]
17. Pang, J.; Feng, C.; Zhu, X.; Liu, B.; Deng, T.; Gao, Y.; Li, Y.; Ke, J. Fracture behaviors of maxillary central incisors with flared root canals restored with CAD/CAM integrated glass fiber post-and-core. *Dent. Mater. J.* **2019**, *38*, 114–119. [CrossRef] [PubMed]
18. Balkenhol, M.; Wöstmann, B.; Rein, C.; Ferger, P. Survival time of cast post and cores: A 10-year retrospective study. *J. Dent.* **2007**, *35*, 50–58. [CrossRef]
19. Heydecke, G.; Butzm, F.; Hussein, A.; Strub, J.R. Fracture strength after dynamic loading of endodontically treated teeth restored with different post-and-core systems. *J. Prosthet. Dent.* **2002**, *87*, 438–445. [CrossRef] [PubMed]
20. Awad, M.A.; Marghalani, T.Y. Fabrication of a custom-made ceramic post and core using CAD-CAM technology. *J. Prosthet. Dent.* **2007**, *98*, 161–162. [CrossRef]
21. Lee, J.H.; Sohn, D.S.; Lee, C.H. Fabricating a fiber-reinforced post and zirconia core with CAD/CAM technology. *J. Prosthet. Dent.* **2014**, *112*, 683–685. [CrossRef] [PubMed]
22. Baba, N.Z.; Golden, G.; Goodacre, C.J. Nonmetallic prefabricated dowels: A review of compositions, properties, laboratory, and clinical test results. *J. Prosthodont.* **2009**, *18*, 527–536. [CrossRef] [PubMed]
23. Ozkurt, Z.; Işeri, U.; Kazazoğlu, E. Zirconia ceramic post systems: A literature review and a case report. *Dent. Mater. J.* **2010**, *29*, 233–245. [CrossRef] [PubMed]
24. Massa, F.; Dias, C.; Blos, C.E. Resistance to fracture of mandibular premolars restored using post and-core systems. *Quintessence Int.* **2010**, *41*, 49–57. [PubMed]
25. Ayad, M.F.; Bahannan, S.A.; Rosenstiel, S.F. Influence of Irrigant, Dowel Type, and RootReinforcing Material on Fracture Resistance of Thin-Walled Endodontically Treated Teeth. *J. Prosthodont.* **2011**, *20*, 180–189. [CrossRef]
26. Alhajj, M.N.; Qi, C.H.; Sayed, M.E.; Johari, Y.; Ariffin, Z. Fracture Resistance of Titanium and Fiber Dental Posts: A Systematic Review and Meta-Analysis. *J. Prosthodont.* **2022**, *31*, 374–384. [CrossRef]

27. Ranjkesh, B.; Leo, M.; Vafaei, A.; Lovschall, H. Pull-Out Bond Strength of Titanium Post Cemented with Novel Fast-Setting Calcium Silicate Cement. *Eur. Endod. J.* **2021**, *6*, 314–318. [CrossRef] [PubMed]
28. *International Standard 28399*; Dentistry—Products for External Tooth Bleaching. International Organization for Standardization: Geneva, Switzerland, 2011.
29. Rho, J.Y.; Ashman, R.B.; Turner, C.H. Young's modulus of trabecular and cortical bone material: Ultrasonic and microtensile measurements. *J. Biomech.* **1993**, *26*, 111–119. [CrossRef]
30. Khaledi, A.A.; Sheykhian, S.; Khodaei, A. Evaluation of Retention of two Different Cast Post-Core Systems and Fracture Resistance of the Restored Teeth. *J. Dent.* **2015**, *16*, 121–128.
31. Sorensen, J.A.; Engelman, M.J. Ferrule design and fracture resistance of endodontically treated teeth. *J. Prosthet. Dent.* **1990**, *63*, 529–536. [CrossRef]
32. Eissman, H.F.; Radke, R.A. Postendodontic restoration. In *Pathways of the Pulp*, 4th ed.; Cohen, S., Burns, R.C., Eds.; Mosby: St. Louis, CV, USA, 1987; pp. 640–643.
33. Bakirtzoglou, E.; Kamalakidis, S.N.; Pissiotis, A.L.; Michalakis, K. In vitro assessment of retention and resistance failure loads of complete coverage restorations made for anterior maxillary teeth restored with two different cast post and core designs. *J. Clin. Exp. Dent.* **2019**, *11*, e225–e230. [CrossRef]
34. De-Deus, G.; Belladonna, F.G.; Silva, E.J.N.L.; Souza, E.M.; Carvalhal, J.C.A.; Perez, R.; Lopes, R.T.; Versiani, M.A. Micro-CT assessment of dentinal micro-cracks after root canal filling procedures. *Int. Endod. J.* **2017**, *50*, 895–901. [CrossRef]
35. Sorensen, J.A.; Martinoff, J.T. Clinically significant factors in dowel design. *J. Prosthet. Dent.* **1984**, *52*, 28–35. [CrossRef]
36. Goodacre, C.J.; Spolnick, K.J. The prosthodontic management of endodontically treated teeth: A literature review. Part II. Maintaining the apical seal. *J. Prosthod.* **1995**, *4*, 51–53. [CrossRef]
37. Morgano, S.M. Restoration of pulpless teeth: Application of traditional principles in present and future contexts. *J. Prosthet. Dent.* **1996**, *75*, 375–380. [CrossRef]
38. Mattison, G.D.; Delivanis, P.D.; Thacker, R.W., Jr.; Hassell, K.J. Effect of post preparation on the apical seal. *J. Prosthet. Dent.* **1984**, *51*, 785–789. [CrossRef]
39. Sreedevi, S.; Sanjeev, R.; Raghavan, R.; Abraham, A.; Rajamani, T.; Govind, G.K. An In Vitro Study on the Effects of Post-Core Design and Ferrule on the Fracture Resistance of Endodontically Treated Maxillary Central Incisors. *J. Int. Oral Health* **2015**, *7*, 37–41. [PubMed]
40. Blatz, M.B.; Chiche, G.; Holst, S.; Sadan, A. Influence of surface treatment and simulated aging on bond strengths of luting agents to zirconia. *Quintessence Int.* **2007**, *38*, 745–753.
41. Amaral, F.L.; Colucci, V.; Palma-Dibb, R.G.; Corona, S.A. Assessment of in vitro methods used to promote adhesive interface degradation: A critical review. *J. Esthet. Restor. Dent.* **2007**, *19*, 340–353. [CrossRef]
42. Stegaroiu, R.; Yamada, H.; Kusakari, H.; Miyakawa, O. Retention and failure mode after cyclic loading in two post and core systems. *J. Prosthet. Dent.* **1996**, *75*, 506–511. [CrossRef]
43. Ahmad, M.; Tarmeze, A.; Abdul Rasib, A. Capability of 3D Printing Technology in Producing Molar Teeth Prototype. *Int. J. Eng. Appl.* **2020**, *8*, 64–70. [CrossRef]
44. Hudis, S.I.; Goldstein, G.R. Restoration of endodontically treated teeth: A review of the literature. *J. Prosthet. Dent.* **1986**, *55*, 33–38. [CrossRef]
45. Bolhuis, H.P.B.; De Gee, A.J.; Feilzer, A.J.; Davidson, C.L. Fracture strength of different core build-up designs. *Am. J. Dent.* **2001**, *14*, 286–290.
46. Liu, W.; Qing, H.; Pei, X.; Wang, J. Internal adaptation of cobalt-chromium posts fabricated by selective laser melting technology. *J. Prosthet. Dent.* **2019**, *121*, 455–460. [CrossRef] [PubMed]
47. Maya, A.; Millsten, P.; Freeman, Y. Determining post-core retention of smooth surface metal, non-metal posts. *J. Dent. Res.* **1998**, *77*, 160.
48. Cohen, B.I.; Pagnillo, M.K.; Newman, I.; Musikant, B.L.; Deutsch, A.S. Retention of three endodontic posts cemented with five dental cements. *J. Prosthet. Dent.* **1998**, *79*, 520–525. [CrossRef]
49. Liberman, R.; Ben-Amar, A.; Urstein, M.; Gontar, G.; Fitzig, S. Conditioning of root canals prior to dowel cementation with composite luting cement and two dentine adhesive systems. *J. Oral Rehabil.* **1989**, *16*, 597–602. [CrossRef] [PubMed]
50. Chapman, K.W.; Worley, J.L.; von Fraunhofer, J.A. Effect of bonding agents on retention of posts. *Gen. Dent.* **1985**, *33*, 128–130. [PubMed]
51. Fernandes, V.; Silva, A.S.; Carvalho, O.; Henriques, B.; Silva, F.S.; Özcan, M.; Souza, J.C.M. The resin-matrix cement layer thickness resultant from the intracanal fitting of teeth root canal posts: An integrative review. *Clin. Oral Investig.* **2021**, *25*, 5595–5612. [CrossRef] [PubMed]
52. Penelas, A.G.; Piedade, V.M.; Borges, A.C.; Poskus, L.T.; da Silva, E.M.; Guimarães, J.G. Can cement film thickness influence bond strength and fracture resistance of fiber reinforced composite posts? *Clin. Oral Investig.* **2016**, *20*, 849–855. [CrossRef]
53. Gomes, G.M.; Rezende, E.C.; Gomes, O.M.; Gomes, J.C.; Loguercio, A.D.; Reis, A. Influence of the resin cement thickness on bond strength and gap formation of fiber posts bonded to root dentin. *J. Adhes. Dent.* **2014**, *16*, 71–88.
54. Amin, R.A.; Mandour, M.H.; Abd El-Ghany, O.S. Fracture strength and nanoleakage of weakened roots reconstructed using relined glass fiber-reinforced dowels combined with a novel prefabricated core system. *J. Prosthodont.* **2014**, *23*, 484–494. [CrossRef] [PubMed]

55. Tamac, E.; Toksavul, S.; Toman, M. Clinical marginal and internal adaptation of CAD/CAM milling, laser sintering, and cast metal ceramic crowns. *J. Prosthet. Dent.* **2014**, *112*, 909–913. [CrossRef] [PubMed]
56. Scotti, N.; Forniglia, A.; Bergantin, E.; Paolino, D.S.; Pasqualini, D.; Berutti, E. Fibre post adaptation and bond strength in oval canals. *Int. Endod. J.* **2014**, *47*, 366–372. [CrossRef] [PubMed]

applied sciences

MDPI

Article

Martens Hardness of CAD/CAM Resin-Based Composites

Martin Rosentritt, Sebastian Hahnel *, Sibylle Schneider-Feyrer, Thomas Strasser and Alois Schmid

Department of Prosthetic Dentistry, UKR University Hospital Regensburg, 93042 Regensburg, Germany; martin.rosentritt@ukr.de (M.R.); sibylle.schneider-feyrer@ukr.de (S.S.-F.); thomas.strasser@ukr.de (T.S.); alois.schmid@ukr.de (A.S.)

* Correspondence: sebastian.hahnel@ukr.de; Tel.: +49-941-944-11900

Abstract: (1) Background: The properties of CAD/CAM resin-based composites differ due to differences in their composition. Instrumented indentation testing can help to analyze these differences with respect to hardness, as well as energy-converting capabilities due to viscoelastic behavior. (2) Methods: Eleven materials were investigated using instrumented indentation testing. Indentation depth (h_r), Martens hardness (H_M), indentation hardness (H_{IT}), indentation modulus (E_{IT}), the elastic part of indentation work (η_{IT}), and indentation creep (C_{IT}) were investigated, and statistical analysis was performed using one-way ANOVA, Bonferroni post-hoc test, and Pearson correlation ($\alpha = 0.05$). (3) Results: All of the investigated parameters revealed differences between the analyzed materials. Besides the differences in hardness-associated parameters (h_r, H_M, and H_{IT}), instrumented indentation testing demonstrated differences in energy-converting properties. The subsequent one-way ANOVA revealed significant differences ($p < 0.001$). A significant ($p < 0.01$, Pearson correlation >0.576) correlation between the materials and H_M, H_{IT}, or E_{IT} was identified. (4) Conclusions: Due to the differences found in the energy-converting properties of the investigated materials, certain CAD/CAM resin-based composites could show superior stress-breaking capabilities than others. The consequential reduction in stress build-up may prove to beneficial, especially for implant-retained restorations or patients suffering from parafunctions.

Keywords: CAD/CAM; resin composite; hardness; instrumented indentation testing

Citation: Rosentritt, M.; Hahnel, S.; Schneider-Feyrer, S.; Strasser, T.; Schmid, A. Martens Hardness of CAD/CAM Resin-Based Composites. *Appl. Sci.* **2022**, *12*, 7698. https://doi.org/10.3390/app12157698

Academic Editor: Mary Anne Melo

Received: 11 July 2022
Accepted: 28 July 2022
Published: 30 July 2022

Publisher's Note: MDPI stays neutral with regard to jurisdictional claims in published maps and institutional affiliations.

1. Introduction

Due to their clearly deviating properties, resin-based CAD/CAM (computer-aided design/computer-aided manufacturing) composites are an interesting clinical alternative to dental ceramics [1]. Similar to direct resin-based composites, resin-based CAD/CAM composites consist of inorganic fillers embedded in an organic polymer matrix, commonly using silanes as coupling agents. Their mechanical properties such as modulus of elasticity or flexural strength are improved due to the standardized polymerization process under industrial conditions compared to chair-side light-curing polymerization [2]. A variation of resin-based composites is the so-called polymer-infiltrated ceramic network (PICN), which comprises a structure-sintered ceramic matrix and a reinforcing polymer network (ceramic content: 86 wt%; polymer content: 14 wt%). Resin-based CAD/CAM composites and resin-infiltrated ceramic networks are used for inlays, onlays, and veneers, as well as tooth- and implant-retained crowns. Some composites are even approved for bridges and for use in patients suffering from bruxism.

One key benefit of these resin-based materials—as advertised by many manufacturers—is the dentine-like modulus of elasticity of approximately 10–30 GPa. Although composites do not reach the high aesthetics of ceramics, they are commonly regarded as less hard and brittle, and they cause less wear and stress in antagonistic teeth [3]. These qualities may be beneficial for the rehabilitation of patients suffering from parafunctions such as bruxism. Energy-dissipation capabilities might also be increased by the utilization of resin-based CAD/CAM composites with a low modulus of elasticity [4–7]. Implant-supported restorations, with their lower tactility and elasticity of the osseointegrated implants, might benefit from less stress build-up during normal mastication. For example, there is evidence for improved implant osseointegration with low-modulus titanium implants [8,9]. This phenomena is mostly attributed to the so-called stress-shielding effect, which is caused by the differences of the elastic moduli between implant and bone. The mismatch leads to an insufficient transfer of force and therefore inadequate stimulation of bone remodeling [10]. It is suggested that the stimulation of bone growth may be enhanced by reducing or adjusting the elastic modulus of the restorative material.

Yet, with respect to the mechanical properties of CAD/CAM resin-based composites, previous research suggests a fairly inhomogeneous class of materials [11]. This is mostly attributed to different types, sizes, and amounts of inorganic fillers (approximately 60–85 wt%), as well as the organic matrix [12]. The significant differences in CAD/CAM resin materials, e.g., flexural strength (150–330 MPa) and modulus of elasticity (10.3–30.0 GPa), may have impacts under clinical conditions. To properly evaluate the available materials and perhaps even choose certain materials for specific clinical indications, detailed information on their mechanical behavior is essential. One method for evaluating elastic and viscoelastic behavior is indentation hardness testing. Surface hardness is defined as the resistance against plastic and therefore permanent deformation by indentation. Hardness is commonly measured with methods such as Vickers, Rockwell or Brinell hardness testing. However, indentation testing encompasses more than just permanent deformation, as elastic or even viscoelastic components can also be determined by the measurement. These properties can be measured using instrumented indentation testing, also called Martens hardness (H_M) testing. H_M is derived from the applied force (F) divided by the indentation surface (A_s), which is a function of the indentation depth (h) (Equation (1)).

$$H_M = \frac{F}{A_S(h)} \tag{1}$$

Furthermore, the constant measurement of force and indentation depth provides a force–indentation depth curve, as well as the fundamentals for additional analysis.

The indentation modulus (E_{IT}), which is determined in the compression mode, is related but not identical to the modulus of elasticity, which is determined in the flexure mode [13]. The elastic part of the indentation (expressed by η_{IT}) could help in the assessment of the use of resin-based CAD/CAM composites for use as stress-breakers for implant-supported restorations. The time-dependent response to the indentation of a viscoelastic material [14] can be expressed as indentation creep (C_{IT}), expressing the relative increase of strain under constant force application, e.g., due to the rearrangement of polymer chains. As the deformation caused by creep is of plastic character, C_{IT} can help to estimate the long-term dimensional and mechanical stability of a material [15–18]. Materials that significantly differ in these properties could therefore be used for different applications.

The hypothesis of this study was that different CAD/CAM resin-composite materials show no similarities regarding indentation depth (h_r), Martens hardness (H_M), indentation hardness (H_{IT}), indentation modulus (E_{IT}), the elastic part of indentation work (η_{IT}), and indentation creep (C_{IT}). The obtained results can help to estimate the energy-conversion behavior and therefore the clinical performance of the significantly different materials under masticatory loads, as well as their stress-breaking capabilities.

2. Materials and Methods

Eleven resin-based CAD/CAM materials (n = 6 per material) were investigated using instrumented indentation testing according to ISO 14577-1 [19]. Table 1 provides an overview over the tested materials, as well as some of their properties (modulus of elasticity E (GPa), filler content wt. (%), and fracture strength FS (MPa)) for the better interpretation of the results of this study. Rectangular specimens (10 × 10 × 2 mm) were produced in CAD/CAM dental milling machine (98 milling blank, Inlab MC X5, Dentsply Sirona, Germany) and polished (1000 grit sandpaper, Tegramin 25, Struers, Germany).

Table 1. Materials, manufacturers, abbreviations (Abbr.), modulus of elasticity (E), filler content (wt.), fracture strength (FS) according to manufacturer's specifications or literature: [a] [20], [b] [21], [c] [22], and [d] [11]). Filler content classification: low-fill ≤ 74% wt. ≤ compact.

Material	Manufacturer	Abbr.	E [GPa]	wt. [%] [d]	FS [MPa]
Cerasmart	GC Corp., Tokyo, JP	CS	12.1 [a]	66.9	231
Brilliant Crios	Coltene Holding AG, Altstätten, CH	BC	10.3	72.0	198
Estelite	Tokuyama Dental, Chiyoda, JP	EL	13.8	72.4	225
Block HC	Shofu Dental GmbH, Ratingen, GER	BL	9.5 [b]	64.1	191
Katana Avencia	Kuraray Noritake, Tokyo, JP	KA	12.4	58.6	190
KZR CAD	Yamakin Co. Ltd., Kochi, JP	KC	10.4	69.0	235
Experimental		EX	20.0	78.3	200
Lava Ultimate	3M Deutschland GmbH, Neuss, GER	LU	12.7	75.4	204 [c]
Grandio bloc	VOCO GmbH, Cuxhaven, GER	GB	18.0	84.5	330
VOCO Experimental	VOCO GmbH, Cuxhaven, GER	VO	n/a	79.7	n/a
Vita Enamic	VITA Zahnfabrik, Bad Säckingen, GER	VE	30.0	85.1	150–160

Testing was carried out with a universal hardness-testing machine (ZwickiLine Z2.5, ZwickRoell, Germany; see Figure 1).

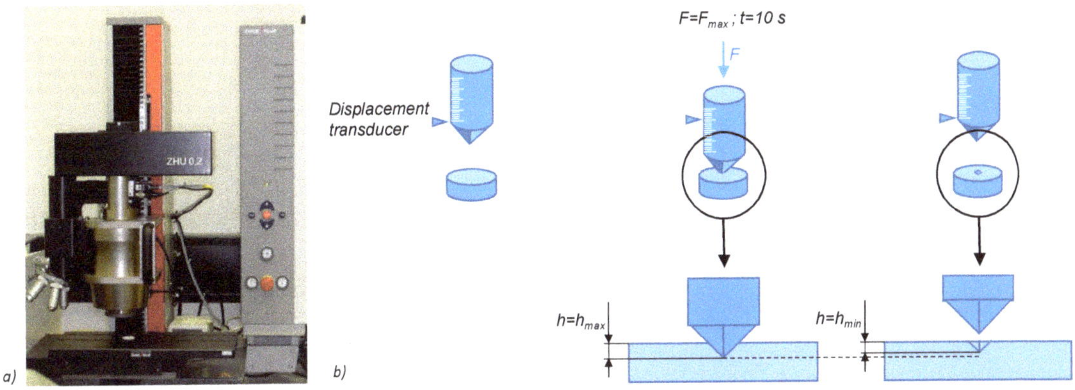

Figure 1. (a) ZwickiLine Z2.5 and (b) schematic test procedure with maximum indentation depth (h_{max}) at application of maximum force (F_{max}) and residual indentation depth (h_{min}) after stress relaxation.

The Martens hardness (HM) is the ratio of the maximum force to the associated contact area (N/mm^2). Other material parameters, such as indentation modulus, indentation creep, and plastic and elastic work of deformation, can be characterized from a force–indentation depth curve. In this study, force, depth and time during the indentation of the diamond pyramid were continuously recorded. The contact area under load was calculated from the maximum indentation depth. The indentation depth was constantly monitored at a loading speed of 0.5 mm/min to a maximum force of F_{max} = 10 N using a Vickers indenter and dwell-time of 10 s. Unloading was performed at 0.1 mm/min. The recorded force–indentation depth curves were used to calculate indentation depth (h_r), Martens hardness (H_M), indentation hardness (H_{IT}), indentation modulus (E_{IT}), the elastic part of indentation work (η_{IT}), and indentation creep (C_{IT}) as defined in ISO 14577-1. The Poisson's ratio of the diamond indenter was set to ν_i = 0.07, and that of the resin-based composite materials was set to ν_s = 0.3 [23]. The Young's modulus of the indenter was E_i = 1140 GPa.

Calculations and statistical analyses were performed using SPSS 25.0 for Windows (IBM, Armonk, NY, USA). The normal distribution of data was controlled using the Shapiro–Wilk test. Means and standard deviations were calculated and analyzed using ANOVA and the Bonferroni test for post-hoc analysis. Pearson correlations were calculated. The level of significance was set to $\alpha = 0.05$.

3. Results

The Shapiro–Wilk test confirmed the normal distribution of the tested parameters. The one-way ANOVA revealed significant differences ($p < 0.001$) within the parameters. Table 2 shows mean results and statistical Bonferroni post-hoc comparison. Force–indentation-curves of the investigated materials are shown in Figure 2.

Table 2. Material, abbreviation (Abbr.), mean and standard deviation (in brackets) for indentation depth (h_r), Martens hardness (H_M), indentation hardness (H_{IT}), indentation modulus (E_{IT}), elastic part of indentation work (η_{IT}), and indentation creep (C_{IT}). Identical superscript letters indicate column-wise non-significant (Bonferroni post hoc test, $p > 0.05$) differences between the materials.

Material	Abbr.	h_r [µm]	H_M [N/mm^2]	H_{IT} [N/mm^2]	E_{IT} [kN/mm^2]	η_{IT} [%]	C_{IT} [%]
Cerasmart	CS	22.8 ^{a,b,d,f} (1.7)	441.3 ^{a,b,d,f} (60.1)	688.5 ^{a,b,c,d,f,g} (88.8)	10.2 ^{a,b,d,f,g,h} (1.9)	48.3 ^{a,b,c,d,f,g,h,i,j} (7.0)	4.7 ^{a,b,c,d,e,f,g} (0.5)
Brilliant Crios	BC	22.7 ^{a,b,d,f} (0.5)	438.5 ^{a,b,d,f} (40.4)	689.5 ^{a,b,c,d,f,g} (40.0)	10.0 ^{a,b,d,f,h} (1.6)	44.1 $^{a–k}$ (2.1)	4.9 ^{a,b,c,d,e,f,g} (0.3)
Estelite	EL	19.4 ^{c,d,e,g,h,j,k} (0.7)	602.8 ^{c,e,g,h,k} (47.9)	940.2 ^{a,b,c,d,e,f,g,h,j,k} (66.8)	13.9 ^{c,f,g} (1.5)	45.2 $^{a–k}$ (1.6)	4.5 ^{a,b,c,d,e,f,g,j} (0.3)
Block HC	BL	22.1 ^{a,b,c,d,f,g} (0.7)	457.3 ^{a,b,d,f,g,h} (27.4)	724.0 ^{a,b,c,d,f,g} (46.1)	10.2 ^{a,b,d,f,g,h} (0.6)	48.5 ^{a,b,c,d,f,g,h,i,j} (0.7)	4.5 ^{a,b,c,d,e,f,g,h,j} (0.2)
Katana Avencia	KA	23.0 ^{a,b,d,f} (1.4)	410.8 ^{a,b,d,f} (55.8)	666.5 ^{a,b,c,d,f,g} (88.2)	8.8 ^{a,b,d,f} (1.4)	50.0 ^{a,b,c,d,f,g,h,i} (2.3)	5.0 ^{a,b,c,d,e,f,g} (0.2)
KZR CAD	KC	19.7 ^{c,d,e,g,j,k} (1.6)	584.0 ^{c,d,e,g,j,k} (45.1)	920.5 ^{a,b,c,d,e,f,g,h,j,k} (141.6)	13.5 ^{a,c,d,g,h} (0.6)	45.6 $^{a–k}$ (7.4)	5.1 ^{a,b,c,d,e,f,g} (0.4)
Experimental	EX	18.7 ^{c,e,g,h,j,k} (0.1)	694.7 ^{c,e,g,h,j,k} (8.5)	1034.5 ^{c,e,g,h,j,k} (12.3)	17.8 ^{e,j,k} (0.3)	40.5 ^{b,c,e,g,j,k} (0.3)	4.8 ^{a,b,c,d,e,f,g} (0.4)
Lava Ultimate	LU	19.1 ^{c,e,g,h,j,k} (3.3)	588.3 ^{c,d,e,g,h,k} (122.6)	1008.7 ^{c,e,g,h,j,k} (331.2)	12.2 ^{a,b,c,d,g,h} (2.2)	50.4 ^{a,b,c,d,f,g,h,i} (4.3)	3.8 ^{d,h,i,j,k} (0.4)
Grandio bloc	GB	17.4 ^{c,e,g,h,j,k} (0.8)	771.2 ^{e,j,k} (73.7)	1184.8 ^{c,e,g,h,j,k} (107.8)	18.4 ^{e,j,k} (2.0)	42.7 ^{a,b,c,d,e,g,i,j,k} (1.5)	4.0 ^{c,d,h,j,k} (0.2)
VOCO Experimental	VO	18.7 ^{c,e,g,h,j,k} (0.3)	692.0 ^{c,e,g,h,j,k} (212.0)	1036.8 ^{c,e,g,h,j,k} (32.3)	17.5 ^{e,j,k} (0.8)	40.2 ^{b,c,e,g,j,k} (0.5)	3.8 ^{h,i,j,k} (0.2)
Vita Enamic	VE	13.9 (0.7)	1143.3 (124.7)	1834.2 (198.0)	25.3 (3.0)	49.2 ^{a,b,c,d,f,g,h,i,j} (0.7)	3.2 ^{h,i,k} (0.1)

The mean Martens hardness (H_M) ranged from 410.8 $\pm$ 55.8 N/mm^2 (KA) to 1143.4 $\pm$ 124.7 N/mm^2 (VE). The ANOVA showed significant ($p \leq 0.001$) differences between the results (Table 2). The residual indentation depth (h_r) was between 13.9 $\pm$ 0.7 µm (VE) and 23.0 $\pm$ 1.4 µm (KA), with significant differences between the results (ANOVA: $p \leq 0.001$). The indentation hardness (H_{IT}) varied between 666.5 $\pm$ 88.2 N/mm^2 (KA) and 1834.2 $\pm$ 198.0 N/mm^2 (VE). The ANOVA confirmed significant ($p \leq 0.001$) differences between the results. The indentation modulus (E_{IT}) ranged from 8.8 $\pm$ 1.4 kN/mm^2 (KA) to 25.3 $\pm$ 3.0 kN/mm^2 (VE), with significant differences between the results (ANOVA: $p \leq 0.001$). The mean elastic part of indentation (η_{IT}) varied between 40.2 $\pm$ 0.5% (VO) and 50.4 $\pm$ 4.3% (LU) (Figure 3). The ANOVA confirmed significant differences between the mean values ($p \leq 0.001$). The indentation creep (C_{IT}) ranged from 3.2 $\pm$ 0.1% (VE) to 5.1 $\pm$ 0.4% (KC). The ANOVA showed significant ($p \leq 0.001$) differences between the values.

A highly significant ($p < 0.01$, Pearson correlation >0.576) correlation between the materials and H_M, H_{IT} or E_{IT} was identified. Negative correlations were established for h_r (−0.623), and C_{IT} (−0.584). No correlation could be determined for η_{IT} (−0.151, $p = 0.227$). A significant ($p < 0.001$) impact of the material was found on H_M ($\eta^2 = 0.914$), H_{IT} ($\eta^2 = 0.867$), E_{IT} ($\eta^2 = 0.910$), C_{IT} ($\eta^2 = 0.771$), η_{IT} ($\eta^2 = 0.544$), and h_r ($\eta^2 = 0.814$).

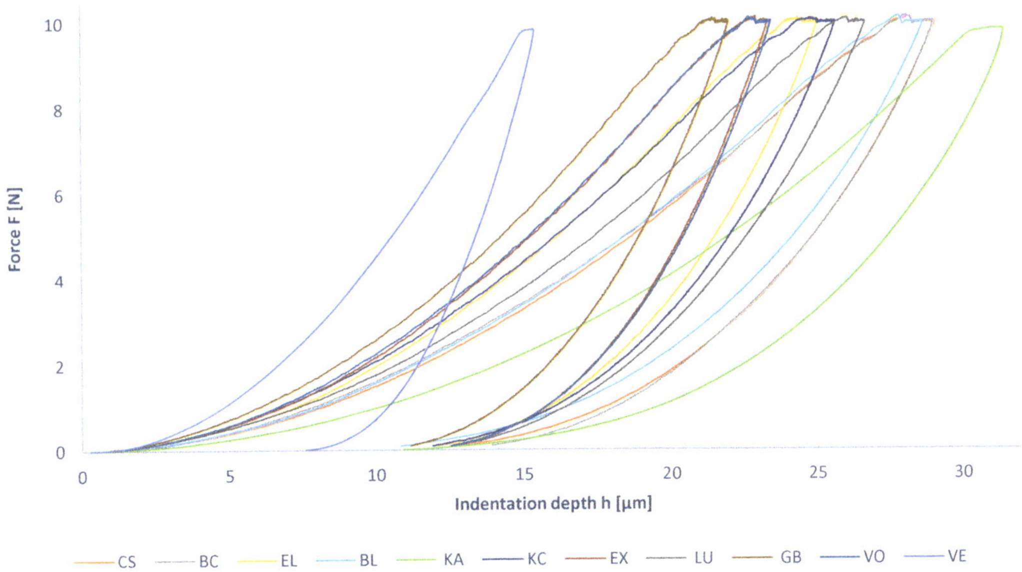

Figure 2. Force–indentation depth curves of the tested composite materials.

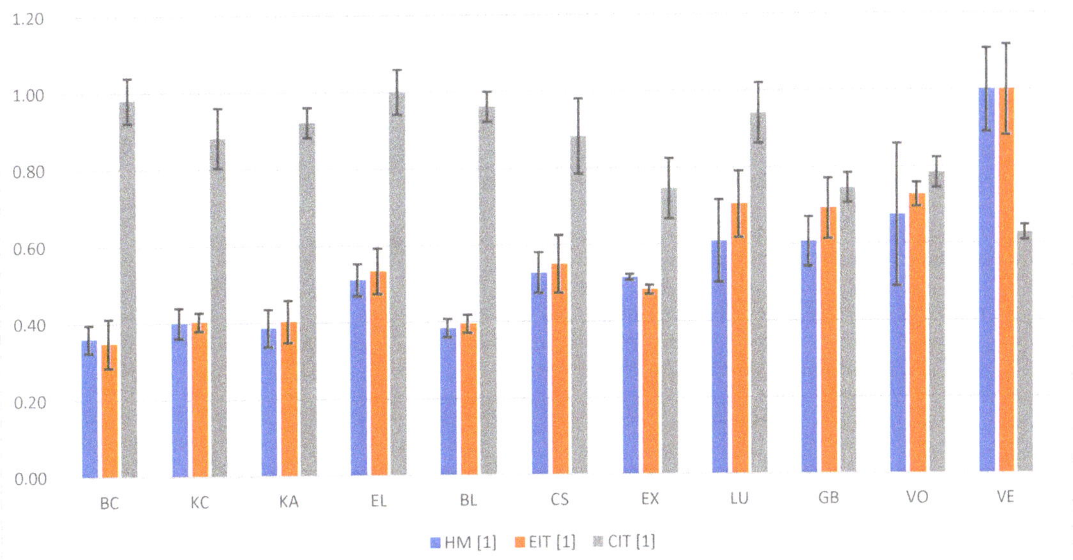

Figure 3. Martens hardness (H_M), indentation modulus (E_{IT}), and indentation creep (C_{IT}) scaled to percentage of maximum value. Materials ordered by increasing amount of inorganic filler content.

4. Discussion

The hypothesis that different CAD/CAM resin-composite materials show no similarities regarding indentation depth (h_r), Martens hardness (H_M), indentation hardness (H_{IT}), indentation modulus (E_{IT}), the elastic part of indentation work (η_{IT}), and indentation

5. Conclusions

The authors of this study investigated the mechanical properties (indentation depth (h_r), Martens hardness (H_M), indentation hardness (H_{IT}), indentation modulus (E_{IT}), elastic part of indentation work (η_{IT}), and indentation creep (C_{IT})) of CAD/CAM resin-based composites with instrumented indentation testing. The Martens hardness and energy-conversion capabilities of eleven different CAD/CAM composites were investigated with a reference to clinical application. Within the limitations of this study, the following conclusions can be drawn:

- The clinical behavior of dental restorations can be influenced by selecting materials based on different elastic and viscoelastic surface parameters.
- Hardness, indentation modulus, and creep vary significantly between different CAD/CAM resin-based composites.
- Individual CAD/CAM resin-composites show different stress-breaking capacities for implant-retained crowns. The reduced resiliency and tactility of implants might be compensated for by a material with low E_{IT} and C_{IT} values.
- Materials with high E_{IT} and low C_{IT} values might be beneficial for partial or inlay crown applications.

Author Contributions: Conceptualization and writing—review and editing, M.R.; writing—review and editing, S.H.; investigation and data curation, S.S.-F.; writing—review and editing, T.S.; writing—original draft preparation, A.S. All authors have read and agreed to the published version of the manuscript.

Funding: This research received no external funding.

Institutional Review Board Statement: Not applicable.

Informed Consent Statement: Not applicable.

Data Availability Statement: Data are contained within the article.

Acknowledgments: We thank the manufacturers for providing the investigated materials.

Conflicts of Interest: The authors declare no conflict of interest.

References

1. Miura, S.; Fujisawa, M. Current status and perspective of CAD/CAM-produced resin composite crowns: A review of clinical effectiveness. *Jpn. Dent. Sci.* **2020**, *56*, 184–189. [CrossRef] [PubMed]
2. Alt, V.; Hannig, M.; Wöstmann, B.; Balkenhol, M. Fracture strength of temporary fixed partial dentures: CAD/CAM versus directly fabricated restorations. *Dent. Mater.* **2011**, *27*, 339–347. [CrossRef] [PubMed]
3. Ruse, N.D.; Sadoun, M.J. Resin-composite Blocks for Dental CAD/CAM Applications. *J. Dent. Res.* **2014**, *93*, 1232–1234. [CrossRef] [PubMed]
4. Skalak, R. Biomechanical considerations in osseointegrated prostheses. *J. Prosthet. Dent.* **1983**, *49*, 843–848. [CrossRef]
5. Rohr, N.; Coldea, A.; Zitzmann, N.U.; Fischer, J. Loading capacity of zirconia implant supported hybrid ceramic crowns. *Dent. Mater.* **2015**, *31*, e279–e288. [CrossRef] [PubMed]
6. Rosentritt, M.; Schneider-Feyrer, S.; Behr, M.; Preis, V. In Vitro Shock Absorption Tests on Implant-Supported Crowns: Influence of Crown Materials and Luting Agents. *Int. J. Oral. Maxillofac. Implants* **2017**, *33*, 116–122. [CrossRef] [PubMed]
7. Niem, T.; Gonschorek, S.; Wöstmann, B. New method to differentiate surface damping behavior and stress absorption capacities of common CAD/CAM restorative materials. *Dent. Mater.* **2021**, *37*, e213–e230. [CrossRef]
8. Shi, L.; Wang, L.; Duan, Y.; Lei, W.; Wang, Z.; Li, J.; Fan, X.; Li, X.; Li, S.; Guo, Z. The Improved Biological Performance of a Novel Low Elastic Modulus Implant. *PLoS ONE* **2013**, *8*, e55015. [CrossRef] [PubMed]
9. Brizuela, A.; Climent, M.; Rios-Carrasco, E.; Rios-Santos, J.; Perez, R.; Manero, J.M.; Gil, F.J. Influence of the Elastic Modulus on the Osseointegration of Dental Implants. *Materials* **2019**, *12*, 980. [CrossRef] [PubMed]
10. Zhang, Y.; Wang, J.; Wang, P.; Fan, X.; Li, X.; Fu, J.; Li, S.; Fan, H.; Guo, Z. Low elastic modulus contributes to the osteointegration of titanium alloy plug. *J. Biomed. Mater. Res. Part. B Appl. Biomater.* **2013**, *101*, 584–590. [CrossRef]
11. Rosentritt, M.; Schneider-Feyrer, S.; Strasser, T.; Koenig, A.; Schmohl, L.; Schmid, A. Thermoanalytical Investigations on the Influence of Storage Time in Water of Resin-Based CAD/CAM Materials. *Biomedicines* **2021**, *9*, 1779. [CrossRef] [PubMed]

12. Rosentritt, M.; Huber, C.; Strasser, T.; Schmid, A. Investigating the mechanical and optical properties of novel Urethandimethacrylate (UDMA) and Urethanmethacrylate (UMA) based rapid prototyping materials. *Dent. Mater.* **2021**, *37*, 1584–1591. [CrossRef] [PubMed]
13. Oliver, W.C.; Pharr, G.M. An improved technique for determining hardness and elastic modulus using load and displacement sensing indentation experiments. *J. Mater. Res.* **1992**, *7*, 1564–1583. [CrossRef]
14. Shahdad, S.A.; McCabe, J.F.; Bull, S.; Rusby, S.; Wassell, R.W. Hardness measured with traditional Vickers and Martens hardness methods. *Dent. Mater.* **2007**, *23*, 1079–1085. [CrossRef] [PubMed]
15. Baroudi, K.; Silikas, N.; Watts, D.C. Time-dependent visco-elastic creep and recovery of flowable composites. *Eur. J. Oral. Sci.* **2007**, *115*, 517–521. [CrossRef]
16. Singh, V.; Misra, A.; Parthasarathy, R.; Ye, Q.; Park, J.; Spencer, P. Mechanical properties of methacrylate-based model dentin adhesives: Effect of loading rate and moisture exposure. *J. Biomed. Mater. Res. B Appl. Biomater.* **2013**, *101*, 1437–1443. [CrossRef] [PubMed]
17. Wang, X.; Zhou, J.; Kang, D.; Swain, M.V.; Menčík, J.; Jian, Y.; Zhao, K. The bulk compressive creep and recovery behavior of human dentine and resin-based dental materials. *Dent. Mater.* **2020**, *36*, 366–376. [CrossRef] [PubMed]
18. Iwasaki, T.; Kamiya, N.; Hirayama, S.; Tanimoto, Y. Evaluation of the mechanical behavior of bulk-fill and conventional flowable resin composites using dynamic micro-indentation. *Dent. Mater. J.* **2022**, *41*, 87–94. [CrossRef]
19. *ISO 14577-1*; 2002: Metallic Materials—Instrumented Indentation Test for Hardness and Materials Parameters—Part 1: Test method. ISO: Geneva, Switzerland, 2002.
20. Wendler, M.; Belli, R.; Petschelt, A.; Mevec, D.; Harrer, W.; Lube, T.; Danzer, R.; Lohbauer, U. Chairside CAD/CAM materials. Part 2: Flexural strength testing. *Dent. Mater.* **2017**, *33*, 99–109. [CrossRef] [PubMed]
21. Lauvahutanon, S.; Takahashi, H.; Shiozawa, M.; Iwasaki, N.; Asakawa, Y.; Oki, M.; Finger, W.J.; Arksornnukit, M. Mechanical properties of composite resin blocks for CAD/CAM. *Dent. Mater. J.* **2014**, *33*, 705–710. [CrossRef]
22. Alamoush, R.A.; Silikas, N.; Salim, N.A.; Al-Nasrawi, S.; Satterthwaite, J.D. Effect of the Composition of CAD/CAM Composite Blocks on Mechanical Properties. *BioMed Res. Int.* **2018**, *2018*, 4893143. [CrossRef] [PubMed]
23. Choi, B.-J.; Yoon, S.; Im, Y.-W.; Lee, J.-H.; Jung, H.-J.; Lee, H.-H. Uniaxial/biaxial flexure strengths and elastic properties of resin-composite block materials for CAD/CAM. *Dent. Mater.* **2019**, *35*, 389–401. [CrossRef] [PubMed]
24. Hampe, R.; Lümkemann, N.; Sener, B.; Stawarczyk, B. The effect of artificial aging on Martens hardness and indentation modulus of different dental CAD/CAM restorative materials. *J. Mech. Behav. Biomed. Mater.* **2018**, *86*, 191–198. [CrossRef] [PubMed]
25. Hampe, R.; Theelke, B.; Lümkemann, N.; Stawarczyk, B. Impact of artificial aging by thermocycling on edge chipping resistance and Martens hardness of different dental CAD-CAM restorative materials. *J. Prosthet. Dent.* **2021**, *125*, 326–333. [CrossRef] [PubMed]
26. Ilie, N. Altering of optical and mechanical properties in high-translucent CAD-CAM resin composites during aging. *J. Dent.* **2019**, *85*, 64–72. [CrossRef] [PubMed]
27. Ilie, N. Spatial Distribution of the Micro-Mechanical Properties in High-Translucent CAD/CAM Resin-Composite Blocks. *Materials* **2020**, *13*, 3352. [CrossRef] [PubMed]
28. Li, Y.; Swartz, M.L.; Phillips, R.W.; Moore, B.K.; Roberts, T.A. Effect of filler content and size on properties of composites. *J. Dent. Res.* **1985**, *64*, 1396–1401. [CrossRef] [PubMed]
29. Kim, K.-H.; Ong, J.L.; Okuno, O. The effect of filler loading and morphology on the mechanical properties of contemporary composites. *J. Prosthet. Dent.* **2002**, *87*, 642–649. [CrossRef] [PubMed]
30. Bociong, K.; Szczesio, A.; Krasowski, M.; Sokolowski, J. The influence of filler amount on selected properties of new experimental resin dental composite. *Open Chem.* **2018**, *16*, 905–911. [CrossRef]
31. Randolph, L.D.; Palin, W.M.; Leloup, G.; Leprince, J.G. Filler characteristics of modern dental resin composites and their influence on physico-mechanical properties. *Dent. Mater.* **2016**, *32*, 1586–1599. [CrossRef] [PubMed]
32. Stöckl, C.; Hampe, R.; Stawarczyk, B.; Haerst, M.; Roos, M. Macro- and microtopographical examination and quantification of CAD-CAM composite resin 2- and 3-body wear. *J. Prosthet. Dent.* **2018**, *120*, 537–545. [CrossRef] [PubMed]
33. Yano, H.T.; Ikeda, H.; Nagamatsu, Y.; Masaki, C.; Hosokawa, R.; Shimizu, H. Correlation between microstructure of CAD/CAM composites and the silanization effect on adhesive bonding. *J. Mech. Behav. Biomed. Mater.* **2020**, *101*, 103441. [CrossRef] [PubMed]
34. Koenig, A.; Schmidtke, J.; Schmohl, L.; Schneider-Feyrer, S.; Rosentritt, M.; Hoelzig, H.; Kloess, G.; Vejjasilpa, K.; Schulz-Siegmund, M.; Fuchs, F.; et al. Characterisation of the Filler Fraction in CAD/CAM Resin-Based Composites. *Materials* **2021**, *14*, 1986. [CrossRef] [PubMed]
35. Halgaš, R.; Dusza, J.; Kaiferova, J.; Kovacsova, L.; Markovska, N. Nanoindentation testing of human enamel and dentin. *Cerami Silik* **2013**, *57*, 92–99.
36. Sadyrin, E.; Swain, M.; Mitrin, B.; Rzhepakovsky, I.; Nikolaev, A.; Irkha, V.; Yogina, D.; Lyanguzov, N.; Maksyukov, S.; Aizikovich, S. Characterization of Enamel and Dentine about a White Spot Lesion: Mechanical Properties, Mineral Density, Microstructure and Molecular Composition. *Nanomaterials* **2020**, *10*, 1889. [CrossRef]
37. Kucher, M.; Dannemann, M.; Modler, N.; Bernhard, M.R.; Hannig, C.; Weber, M.T. Mapping of the Micro-Mechanical Properties of Human Root Dentin by Means of Microindentation. *Materials* **2021**, *14*, 505. [CrossRef] [PubMed]
38. Rees, J.S.; Jacobsen, P.H. The elastic moduli of enamel and dentine. *Clinic. Mater.* **1993**, *14*, 35–39. [CrossRef]

39. Güth, J.F.; Edelhoff, D.; Goldberg, J.; Magne, P. CAD/CAM Polymer vs Direct Composite Resin Core Buildups for Endodontically Treated Molars Without Ferrule. *Oper. Dent.* **2016**, *41*, 53–63. [CrossRef]
40. Bates, J.F.; Stafford, G.D.; Harrison, A. Masticatory function—A review of the literature. 1. The form of the masticatory cycle. *J. Oral. Rehabil.* **1975**, *2*, 281–301. [CrossRef]
41. Duan, Y.; Griggs, J.A. Effect of elasticity on stress distribution in CAD/CAM dental crowns: Glass ceramic vs. polymer–matrix composite. *J. Dent.* **2015**, *43*, 742–749. [CrossRef] [PubMed]
42. Zheng, Z.; He, Y.; Ruan, W.; Ling, Z.; Zheng, C.; Gai, Y.; Yan, W. Biomechanical behavior of endocrown restorations with different CAD-CAM materials: A 3D finite element and in vitro analysis. *J. Prosthet. Dent.* **2021**, *125*, 890–899. [CrossRef] [PubMed]
43. Dejak, B.; Mlotkowski, A.; Langot, C. Three-dimensional finite element analysis of molars with thin-walled prosthetic crowns made of various materials. *Dent. Mater.* **2012**, *28*, 433–441. [CrossRef] [PubMed]
44. Costa, A.K.F.; Xavier, T.A.; Noritomi, P.Y.; Saavedra, G.; Borges, A.L.S. The Influence of Elastic Modulus of Inlay Materials on Stress Distribution and Fracture of Premolars. *Oper. Dent.* **2014**, *39*, E160–E170. [CrossRef] [PubMed]
45. Özkir, S.E. Effect of restoration material on stress distribution on partial crowns: A 3D finite element analysis. *J. Dent. Sci.* **2018**, *13*, 311–317. [CrossRef] [PubMed]

applied sciences

MDPI

Article

Linear Dimensional Change in Acrylic Denture Teeth Positions Factored by Different Processing Techniques and Occlusal Forms: An In Vitro Study

Mohammed E. Sayed [1,*], Amit Porwal [1], Saurabh Jain [1,*], Abdulkarim Hussain Alshehri [1], Nasser M. Alqahtani [2], Ashwaq Hadi Ali Hadadi [3], Rawan Abdulwadoud Zakri [3], Sahar Mahdi Musa Zeed [3], Saswan Ibrahim Nahari [3], Fatimah H. Alsurayyie [1], Hatem Alqarni [4], Saeed M. Alqahtani [2] and Saad Saleh AlResayes [5]

1 Department of Prosthetic Dental Sciences, College of Dentistry, Jazan University, Jazan 45142, Saudi Arabia; aporwal2000@gmail.com (A.P.); dr.alshehri11@gmail.com (A.H.A.); falsoraiee@jazanu.edu.sa (F.H.A.)
2 Department of Prosthodontics, College of Dentistry, King Khalid University, Abha 62529, Saudi Arabia; nmalqahtani@kku.edu.sa (N.M.A.); smaalqahtani@kku.edu.sa (S.M.A.)
3 Private Dental Practice, Jazan 45142, Saudi Arabia; alees007@outlook.sa (A.H.A.H.); rnoozak@gmail.com (R.A.Z.); sahar.9007@outlook.sa (S.M.M.Z.); sawsan7673@gmail.com (S.I.N.)
4 Restorative and Prosthetic Dental Science Department, College of Dentistry, King Abdullah International Medical Research Center, King Saud Bin Abdulaziz University for Health Science, Riyadh 11545, Saudi Arabia; qarnih@ksau-hs.edu.sa
5 Department of Prosthetic Dental Sciences, College of Dentistry, King Saud University, Riyadh 11545, Saudi Arabia; salresayes@ksu.edu.sa
* Correspondence: drsayed203@gmail.com (M.E.S.); drsaurabhjain79@gmail.com (S.J.)

Citation: Sayed, M.E.; Porwal, A.; Jain, S.; Alshehri, A.H.; Alqahtani, N.M.; Hadadi, A.H.A.; Zakri, R.A.; Zeed, S.M.M.; Nahari, S.I.; Alsurayyie, F.H.; et al. Linear Dimensional Change in Acrylic Denture Teeth Positions Factored by Different Processing Techniques and Occlusal Forms: An In Vitro Study. *Appl. Sci.* **2022**, *12*, 7058. https://doi.org/10.3390/app12147058

Academic Editor: Mary Anne Melo

Received: 18 June 2022
Accepted: 11 July 2022
Published: 13 July 2022

Abstract: The current literature lacks substantial evidence for the effect of denture base processing techniques and posterior denture tooth forms on denture tooth shifts due to denture base resin polymerization. The aim of this study was to evaluate the combined effect of PMMA-based denture processing techniques (compression packing and injection molding) and posterior tooth forms (semi-anatomic and non-anatomic) on the linear dimensional shift of denture teeth following denture processing in both horizontal and vertical dimensions. Two different complete denture fabrication techniques were used to prepare forty ideal maxillary complete dentures using two different types of posterior tooth forms. The used fabrication techniques were conventional heat polymerized compression packing and injection molding. The posterior tooth forms used in the current study were non-anatomic tooth (0 degrees) and semi-anatomic tooth forms (approximately 20 degrees). Initial linear measurements (vertical and horizontal) were taken from pre-specified points for the central incisor and first molar. Specimens were randomly divided into four groups ($n = 10$), and denture processing was performed using the two techniques. Final linear measurements were recorded. The linear change in dimension for all six parameters was calculated by deducting the after values from the before values. Since the discrepancies were both positive and negative in magnitude, the absolute value of the difference was taken for further analysis. This value represents the dimensional change. *T*-tests were used to compare the mean dimensional changes. Furthermore, the mean dimensional changes for all the six parameters were compared using a two-way analysis of variance. The alpha error was set at 5%, and a p-value of less than 0.05 was considered statistically significant. The injection molding technique showed significantly fewer tooth movements in both the vertical and horizontal measurements as compared to the conventional compression packing technique. The non-anatomic tooth showed significantly fewer changes in tooth movement as compared to semi-anatomic teeth in both the compression and injection techniques. This study can guide the selection of a proper processing technique for a particular posterior tooth form, thus minimizing occlusal discrepancies and reducing occlusal corrections during laboratory and clinical remount procedures.

Keywords: PMMA; acrylic teeth; non-anatomic teeth; semi-anatomic teeth; injection molding technique; compression molding technique; denture processing; tooth movement; denture base resin; posterior tooth form; linear dimensional change; complete denture

---

1. Introduction

Complete edentulism, as defined by a loss of all permanent dentition, is a global health condition that was reported to affect a range of 7–69% of adult populations worldwide [1]. Even though implant dentistry has changed the approach of treatment planning for completely edentulous cases, conventional complete dentures remain a valid and sometimes first option, especially when implants are contraindicated in such cases. Balanced complete denture occlusion is crucial for the stability and retention of complete dentures in patients with moderate to severe alveolar ridge resorption [2]. Excessive denture tooth movements due to processing may compromise the form, esthetics and occlusion and, therefore, mean that dentures need to be remade [3]. Several factors have been reported to affect the positional stability of a denture tooth during processing. These factors include the investment method [4], investment material [5], flasking and polymerization technique [6,7], denture base thickness [8], flask closure and cooling method [9], palatal form [10] and box preparation on the heel of the master cast [11]. In addition, Goodacre et al. [12] studied the effect of the denture processing technique on denture tooth movement and found that CAD–CAM technology was the most accurate and reproducible in comparison to conventional packing and injection techniques. The study also reported that the packing technique resulted in higher vertical tooth movement. However, CAD–CAM technology for denture fabrication was only implemented in 10% or less of the cases treated in US dental schools [13]. Therefore, the focus should be directed toward perfecting the techniques that are commonly used in complete denture practice. Jackson and colleagues [14] reported that the presence of acrylic teeth has a significant effect on the accuracy of denture bases fabricated using compression and injection techniques. Several studies have indicated that the injection technique is more accurate with respect to changing vertical dimension and occlusal errors [15,16]. The case-specific selection approach of the posterior tooth form is important for maintaining masticatory efficiency and the success of denture treatment [17]. The available evidence is lacking with regard to the effect of posterior denture tooth form (i.e., cuspal height) on denture tooth movement due to acrylic resin processing. Hence, the goal of this study was to evaluate the combined effect of processing techniques (compression and injection molding) and posterior tooth forms (semi-anatomic and non-anatomic) on linear dimension changes (horizontal and vertical dimensions) in denture tooth positions following processing. The null hypothesis of the study was that the variation in posterior tooth forms and complete denture processing technique has no effect on the dimensional position of anterior and posterior teeth following processing.

2. Materials and Methods

The research protocol was approved by the research board at the College of Dentistry, Jazan University (reference number: CODJU-2003I)

2.1. Materials

In this study, two different complete denture fabrication techniques were used to prepare forty ideal maxillary dentures using two different types of posterior tooth forms. The fabrication techniques used were: conventional heat polymerized compression molding and injection molding. The posterior tooth forms used in the current study were non-anatomic tooth (0 degrees) and semi-anatomic tooth forms (approx. 20 degree). Details of the materials used in the fabrication of maxillary complete dentures are listed in Table 1.

Table 1. Materials used in the study.

Denture Base Resin			
Trade Name	**Manufacturer**	**Main Composition**	**Fabrication Technique**
Meliodent, Heraeus Kulzer	Kulzer GmbH, Hanau, Germany	Polymethylmethacrylate	Compression molding, heat polymerizing technique
Breflex Polyan IC, Bredent, Germany	Bredent GmbH & Co. KG, Senden, Germany	Polymethylmethacrylate	Injection molding technique (Thermopress 400 system 2.62)
Acrylic Teeth			
Trade Name	**Tooth Form**	**Manufacturer**	**Main Composition**
Acrylic TruSmile teeth set	Non-anatomic and semi-anatomic	Jayna Industries, Ghaziabad, Uttar Pradesh, India	Prepolymerized PMMA resin
Other Materials Used in the Study			
Material Name	**Trade Name**	**Manufacturer**	**Main Composition**
Dental stone	Durguix	Protechno, Vilamalla, Spain	Alpha hemihydrate
Modelling wax	Spezial sculpturing wax	YETI Dentalprodukte GmbH, Engen, Germany	Paraffin wax
Duplicating silicone	Precisil duplicating silicone	YETI Dentalprodukte GmbH, Engen, Germany	Additional curing silicone (Vinylsiloxane, Platincomplexes and Siliciumdioxid)

2.2. Pouring of the Cast and the Arrangement of Teeth

To standardize the reference points for measurements, a round bur (# 001/027, Quattro, DFS-Diamon GmbH, Ländenstraße, Riedenburg, Germany) was used to make holes on the land area of the ideal edentulous maxillary cast, corresponding to the disto-buccal cusps of the right and left first molars and maxillary right central incisor. The silicone mold made using this reference cast was used to pour forty ideal casts with grooves in the land area (Figure 1).

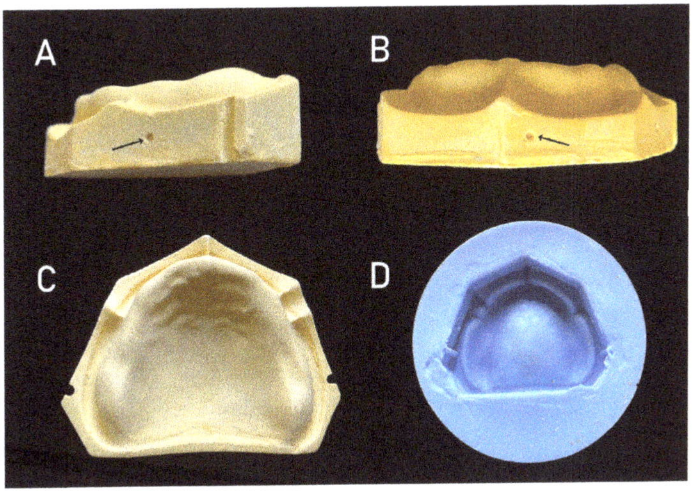

Figure 1. Steps of duplicating cast and teeth setup: (**A**,**B**) round holes made on cast corresponding to maxillary right central incisor and disto-buccal cusps of the right and left first molars; (**C**) ideal edentulous cast with orientation grooves; (**D**) custom edentulous silicone mold.

Two sheets of modelling wax were adapted on the cast, and ideal teeth arrangements for complete dentures were performed. The arrangements of the teeth on all the casts were obtained by the duplication of the ideal teeth arrangement from the initial cast, i.e., twenty sets of semi-anatomical teeth and twenty non-anatomic teeth. Following this, vertical grooves were placed on the right central incisor incisal edge (tooth number 1.1) (bur (# 114/023, Quattro, DFS-Diamon GmbH, Ländenstraße, Riedenburg, Germany)) and the round holes on the disto-buccal cusps or right and left first molars (tooth number 1.6 and 2.6) (bur (# 001/023, Quattro, DFS-Diamon GmbH, Ländenstraße, Riedenburg, Germany)) (Figure 2).

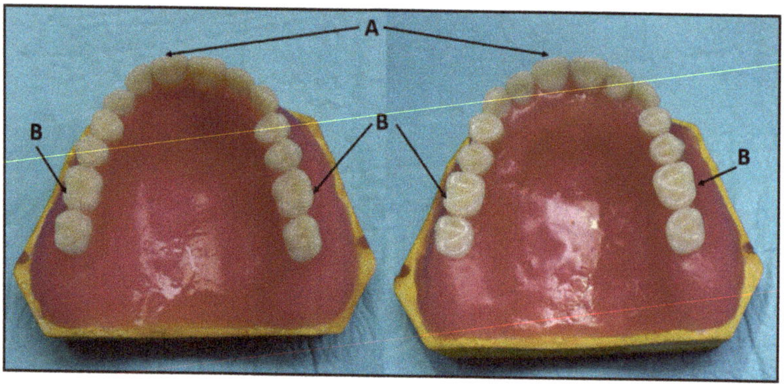

Figure 2. Dentures before processing: (**A**) vertical groove placed on the right central incisor incisal edge; (**B**) round holes on the disto-buccal cusps or right and left first molars.

2.3. *Initial Measurements*

Linear measurements were taken for all the sets of semi-anatomical and non-anatomical teeth. Vertical measurements were taken between the mesio-buccal grooves of the first molar on both the right and left sides, and a round hole was placed on the land area of the cast and also between the groove placed on the right central incisal edge and the round hole on the corresponding land area of the cast [11]. Furthermore, horizontal measurements were taken between the groove on the labial surface of the right central incisor and the holes placed on the distobuccal cusps of the right and left first molars with the heel area of the casts. A metallic ruler was snugly fitted on the heel area of the casts to facilitate these horizontal measurements [11]. Measurements were taken three times for each tooth position and were then averaged.

2.4. *Denture Processing*

PMMA denture base resin was selected for the acrylization of the dentures using two different techniques, namely, the conventional compression molding technique and the injection molding technique. Following measurements, all forty teeth arrangements were randomly divided into 4 groups (n = 10) and were processed by conventional compression molding and injection molding techniques.

G1: Conventional compression molding technique with non-anatomical tooth form;
G2: Injection molding technique with non-anatomical tooth form;
G3: Conventional compression molding technique with semi-anatomical tooth form;
G4: Injection molding technique with semi-anatomical tooth form.

A standardized procedure was followed for the processing of all the dentures according to the manufacturer's instructions. Conventional metal denture flasks were used for the compression molding technique, and a long curing cycle (74 °C for 8 h followed by 100 °C for 1 h) was used to polymerize the resin. For the injection molding technique, special metallic flasks were used. A Thermopress 400 system 2.62 (2019, Bredent GmbH & Co., Senden, Germany) was used. The equipment was programmed at a temperature setting of

265 °C, a heating time of 15 min, an injecting pressure of 9.5 bar and a pressing time of 60 s. After processing, the dentures were left to cool completely and were not retrieved from their respective casts (Figure 3).

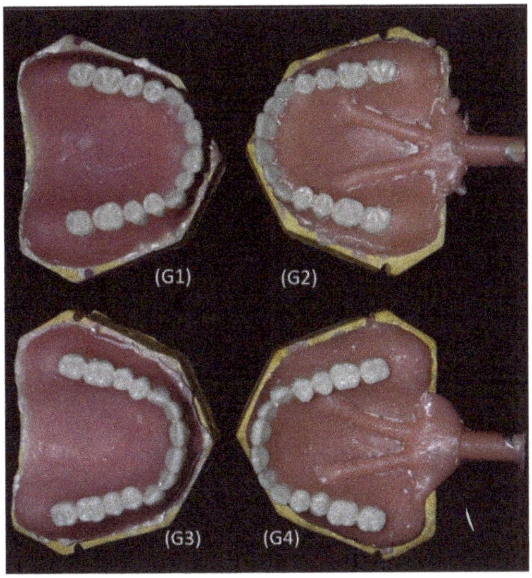

Figure 3. Dentures after processing: G1: conventional compression molding technique with non-anatomical tooth form; G2: injection molding technique with non-anatomical tooth form; G3: conventional compression molding technique with semi-anatomical tooth form, G4: injection molding technique with semi-anatomical tooth form.

2.5. Final Measurements

All the plaster over the surface was cleaned for each specimen, and their measurements (post-processing) were performed at exactly the same positions for all the samples with the help of the Vernier caliper (SE 784EC Digital Caliper; SE Tools, Lapeer, MI, USA) with an accuracy of 0.01 mm. All the initial and final measurements were performed by a single trained investigator. Measurements were taken three times in each dimension and were then averaged to obtain the final reading.

2.6. Data Analysis

The collected data were tabulated in a Microsoft Excel spreadsheet (Microsoft Inc., Redmond, WA, USA), and statistical analysis was performed using SPSS version 24.0 (IBM Corp. Released 2016; IBM SPSS Statistics for Windows, Version 24.0. Armonk, NY, USA: IBM Corp.). The changes in the dimensions for all six parameters were calculated by deducting the after values from the before values. These data were mostly normally distributed, as checked by the Shapiro–Wilk Test (Supplementary Table S1). Since the discrepancies were both positive and negative in magnitude, the absolute value of the difference was taken for further analysis. This value represents the dimensional change. These data showed some deviation from a normal distribution, as shown in Supplementary Table S2. *T*-tests were used to compare the mean dimensional changes. Furthermore, the mean dimensional changes for all the six parameters were compared using a two-way analysis of variance. The alpha error was set at 5%, and a *p*-value of less than 0.05 was considered statistically significant.

3. Results

The mean linear dimensional changes among the groups are presented in Table 2 and Figure 4. The maximum linear dimensional changes in the positions of the molar in the vertical direction were observed in complete dentures with semi-anatomic teeth and that were processed using the conventional compression molding technique (0.27 ± 0.46 and 0.24 ± 0.61). The smallest changes were observed in complete dentures with non-anatomic teeth and that were processed using the injection molding technique (0.02 ± 0.59 and 0.017 ± 0.75). The maximum linear dimensional changes in the positions of the incisor in the vertical direction were observed in complete dentures with semi-anatomic teeth and that were processed using the conventional compression molding technique (0.21 ± 0.89), whereas complete dentures with non-anatomic teeth that were processed using the injection molding technique showed the smallest changes (0.03 ± 0.44).

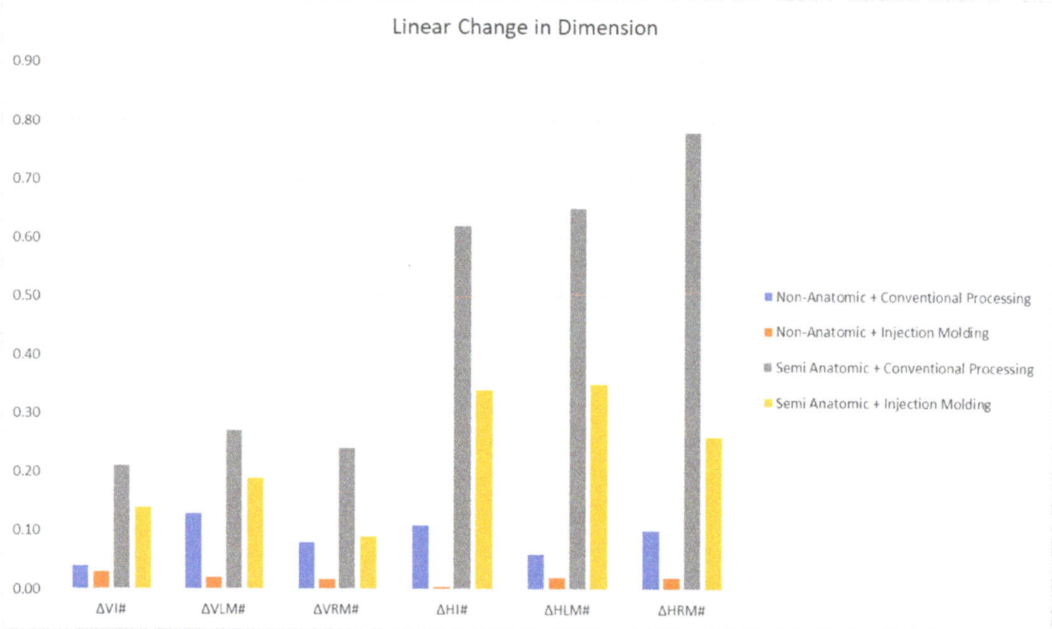

Figure 4. Linear change in dimension of non-anatomic and semi-anatomic processing groups. # Difference calculated by subtracting the post-processing values from the pre-processing values; ΔVI, mean linear dimensional change in the position of the incisor in the vertical direction; ΔVLM, mean linear dimensional change in the position of the left molar in the vertical direction; ΔVRM, mean linear dimensional change in the position of the right molar in the vertical direction; ΔHI, mean linear dimensional change in the position of the incisor in the horizontal direction; ΔHLM, mean linear dimensional change in the position of the left molar in the horizontal direction; ΔHRM, mean linear dimensional change in the position of the right molar in the horizontal direction.

Table 2. Mean, standard deviation (linear change in dimension in mm) and intergroup comparison with respect to non-anatomic and semi-anatomic processing groups.

	Group	n	ΔVI #	ΔVLM #	ΔVRM #	ΔHI #	ΔHLM #	ΔHRM #
Non-anatomic teeth	Conventional processing technique (G1)	10	0.04 ± 0.88	0.13 ± 1.31	0.08 ± 0.57	0.11 ± 0.33	0.06 ± 0.43	0.10 ± 0.34
	Injection molding technique (G2)	10	0.03 ± 0.44	0.02 ± 0.59	0.017 ± 0.75	0.004 ± 0.32	0.02 ± 0.49	0.02 ± 0.54
	p-value	-	0.56	0.003 *	0.055	0.0001 *	0.63	0.004 *
Semi-anatomic teeth	Conventional processing technique (G3)	10	0.21 ± 0.89	0.27 ± 0.46	0.24 ± 0.61	0.62 ± 0.46	0.65 ± 0.79	0.78 ± 0.63
	Injection molding technique (G4)	10	0.14 ± 0.33	0.19 ± 0.65	0.09 ± 0.78	0.34 ± 0.57	0.35 ± 0.64	0.26 ± 0.50
	p-value	-	0.04	0.02	0.001 *	0.0001 *	0.0001 *	0.0001 *

Difference calculated by subtracting the post-processing values from the pre-processing values; * significant ($p < 0.005$); ΔVI, mean linear dimensional change in the position of the incisor in the vertical direction; ΔVLM, mean linear dimensional change in the position of the left molar in the vertical direction; ΔVRM, mean linear dimensional change in the position of the right molar in the vertical direction; ΔHI, mean linear dimensional change in the position of the incisor in the horizontal direction; ΔHLM, mean linear dimensional change in the position of the left molar in the horizontal direction; ΔHRM, mean linear dimensional change in the position of the right molar in the horizontal direction.

When the linear dimensional changes in the positions of the molar in the horizontal direction were assessed, it was observed that complete dentures with semi-anatomic teeth that were processed using the conventional compression molding technique showed the maximum movement (0.65 ± 0.79 and 0.78 ± 0.63). The least movement was observed in complete dentures with non-anatomic teeth that were processed using the injection molding technique (0.02 ± 0.49 and 0.02 ± 0.54). The maximum linear dimensional changes in the positions of the incisor in the horizontal direction were observed in complete dentures with semi-anatomic teeth that were processed using the conventional compression molding technique (0.62 ± 0.46), whereas complete dentures with non-anatomic teeth that were processed using the injection molding technique showed the smallest changes (0.004 ± 0.32).

Table 3 depicts the two-way ANOVA comparison among the four groups with respect to dimensional change. The dimensional change was found to be greater in the conventional technique as compared to the injection molding technique. Additionally, there was a statistically significant difference in the non-anatomic teeth as compared to the semi-anatomic teeth.

Table 3. Two-way ANOVA comparing the tooth forms and processing techniques.

Source		Type III Sum of Squares	df	Mean Square	F	Sig.
Intercept	Hypothesis	26.045	1	26.045	25.969	0.001
	Error	9.372	8.940	1.048 [a]		
Tooth form	Hypothesis	17.056	1	17.056	24.104	0.000
	Error	117.437	176	0.660 [b]		
Processing technique	Hypothesis	5.908	2	2.954	5.689	0.001
	Error	117.437	176	0.660 [b]		

Dependent variable: dimensional change; [a] 167 MS (technique) + 0.833 MS (error); [b] MS (error).

4. Discussion

With respect to their impact on dimensional changes, there was a significant interaction between the components, i.e., the posterior tooth forms and processing techniques applied. The statistical analysis revealed that the results were rather pronounced, with the injection molding technique exhibiting fewer changes during the transition between the waxing stage and post-processing. Another observed pattern was that the non-anatomic teeth positional change was less than that for the semi-anatomic teeth, regardless of the dimension examined and the processing technique used. The outcomes of the study revealed that changes in the positions of the molar were the highest in the horizontal dimension as compared to other dimensions tested for all the groups, when semi-anatomic teeth were processed using the conventional molding technique. However, the smallest differences were observed in the non-anatomic group, when the injection processing technique was used. The significant positional changes in semi-anatomic teeth can be explained by the complex interlock between the denture teeth and the investment material, which may create excessive stresses during acrylic polymerization that lead to significant teeth movement. The results of this study negate the proposed null hypothesis.

In the annals of contemporary dentistry, the introduction of acrylic resins in 1937 was a watershed moment. Their adoption and acknowledgment in prosthodontics are quite remarkable, as they are proven to be more esthetic and simpler to manipulate to be used in the laboratory as well as in clinics [18]. Methacrylates, particularly methyl methacrylate (MMA) acrylic resin, have been used in dentistry as a denture foundation material since their inception and have been widely employed, since they are regarded as suitable substances for use in the oral environment [19,20]. Recently, the usage of pour-type (fluid) resins in the manufacturing of denture bases has expanded dramatically [21].

The dimensional shift that happens during polymerization shrinkage is crucial to the preservation and stability of the complete denture [22]. The movement of teeth occurs during and after the production of complete dentures, according to research published in the literature [9,23]. A careful understanding and consideration of this may allow for the creation of functioning complete dentures, which require fewer occlusal modifications on the articulator and minimal corrections in the patient's mouth. Previous research has found that the highest amount of tooth movement occurs in the posterior teeth [24]; however, this study found that the maximal teeth movement occurs solely in the compression molding semi-anatomic teeth group for this study and also for the molar teeth. Final measurements were performed after the dentures were recovered from flasking but without removal from the casts. The observed dimensional changes were greater, as the dentures were not immersed in water, as suggested previously in the literature [24]. The dimensional changes produced by water sorption cause expansion, presumably due to the entrance of water between polymethyl methacrylate molecules [25]. The movement of teeth during the production of a complete denture is also influenced by a variety of other factors. The influence of base thickness [8], geometric palatal form [10] and closing flask pressure [26] has been studied. The flask-closing procedure may cause tooth displacements, and the post-pressing time association was partially acknowledged. Denture imperfections, including base distortion and false tooth displacement, were addressed through meticulous measurements in the current study.

The literature includes many studies [16,27–29] related to the comparison of the conventional and injection molding processing techniques. Although, in recent decades, the conventional technique has shown promising results and is still widely used, there is still continuous ongoing research related to the development of other processing techniques. The recently introduced injection molding technique has shown promising results, with a minimal incisal pin opening and reduced adjustments required later. However, the movement of the teeth was in both the horizontal and vertical directions; it showed significant differences in the vertical dimension when the compression molding technique was used with PMMA as the material in both the techniques [16]. Bahra et al. [30] studied both linear and volumetric dimensional changes in six injection molding PMMA denture base resins and found that the IvoBase system was more accurate as compared to other resins processed by other curing techniques. Strohaver [29] compared compression and injection molded complete dentures and found fewer vertical dimension changes in the injection molding technique. When a comparison was conducted for the accuracy of the linear measurements of chemically distinct injection molding materials (PMMA, nylon and styrene) to that of compression molding acrylic resin (PMMA) by Parvizi et al. [27], they concluded that nylon had the greatest overall distortion and that styrene had the smallest. According to Sykora [28], the injection molding technique's higher dimensional accuracy compared to the conventional method may be due to smaller resin particles, lower polymerization temperature, the lack of resin layer formation between the flask counterparts and the lack of the movement of the two halves of the flask during resin packing.

When lingualized balanced occlusion was compared with conventional balanced occlusion, the increase in the vertical dimension was found to be similar [31]. In 2004, Parvizi [27] performed a study on dimensional changes between compression and injection molding techniques using different denture base materials. All the materials used in the study responded differently with different dimensional changes in antero-posterior and cross arch measurements. The results of the current study are also in accordance with this study, showing different changes in both the horizontal and vertical dimensions for both the type of teeth and the processing technique used. However, in the complete dentures with non-anatomic teeth that were processed using injection molding technology, the lowest linear dimensional changes in the incisal and molar positions were found, whereas the highest dimensional changes were found in the semi-anatomic teeth with the conventional processing technique.

Venus [32] compared the denture base resins, processing methods and the denture bases with or without teeth and concluded that the processing technique is a more dominant variable than denture base resins. Peyton and Craig [33] reported difficulty in controlling the changes in the vertical dimension and identified pressure during flask closure as the most significant factor. Other mentioned factors that may be related are due to varied temperatures of the flask, dough consistency, stone mold strength and other factors. Additionally, Zakhari mentioned that teeth are under considerable pressure during packing and curing in the conventional curing technique [34]. Shippee [35] concluded from his experiment that, when the volume of acrylic used during final closure is increased, the teeth are more likely to be intruded; however, it can be reduced if proper venting is provided for acrylic. Atkinson and Grant (1962) [36] found in their investigation that the movement of the teeth is mainly due to a change in the dimension of the mold and also due to the pressure of the acrylic on the teeth and mold. They also proposed that teeth in the mold are carried to a different position due to the reconstitution of the gypsum on heating, cooling and compression. Additionally, the short curing technique has shown an increase in the vertical dimension and a lower number of occlusal contacts [37]. However, in the present study, utmost precaution was taken to maintain the same amount of acrylic used for each denture, standardized pressure during flask closure and a long curing cycle to reduce the effect of these factors on the movement of teeth. However, in the injection molding processing technique, the continuous flow of the non-polymerized material from the reservoir sprue would have compensated for the polymerization shrinkage, as also mentioned by Anderson [38].

The literature has shown that maximum changes after polymerization and the release of internal strains occur after polishing and water storage for 24 h. Additionally, it was shown by Venus [32] that intermolar width increases after water storage and decreases by shrinkage, suggesting that the fit of the denture improves after storage in water. Additionally, Chuchulska [39] studied four thermoplastic injection molding materials (Bre.flex 2nd edition, Vertex ThermoSens, Perflex Biosens and Polyan IC) and concluded that shrinkage is present in all the materials tested for the injection molding technique and that storage in water affects all the materials. In the current study, the post-processing measurements were performed after the removal of the dentures from the flasks but without removal from the casts and without any storage in water, which may be attributed to the more dimensional changes that were recorded. Complete dentures with non-anatomic teeth that were manufactured using the injection molding technique showed the smallest changes. However, dentures with semi-anatomic teeth that were processed using the conventional compression molding technique showed the greatest linear dimensional changes in the incisor and molar positions in both the horizontal and vertical directions in the present study.

The prevailing literature lacks studies related to the effects of posterior tooth forms and processing techniques on linear dimensional changes in the teeth. Zakhari [34] used non-anatomic acrylic teeth and established a combination of plaster and stone to reduce the change in the vertical dimension. However, this study also suggested stone with a lower water–powder ratio as an investing medium in order to control expanding acrylic resin. Woelfel et al. [40] compared different denture base materials in their study and noted that the sharper defined anatomy of Pilkington–Turner teeth seemed to coincide with more tooth movement. Carr et al. compared 0-degree teeth and 33-degree teeth and found less vertical tooth movement in the 0-degree teeth as compared to the 33-degree teeth, when plaster of Paris was used as compared to stone as an investing medium [5]. They attributed this to the greater setting expansion of plaster of Paris pushing the teeth into the wax followed by intrusion back in the mold due to the hydraulic pressure of the acrylic. In the current study, non-anatomic teeth also showed less tooth movement in both the vertical and horizontal dimensions as compared to the semi-anatomic teeth. This may be attributed to the sharper occlusal anatomy of the semi-anatomic teeth as compared to the uniform occlusal surface of the non-anatomic teeth, and it may also be related to the pressure during acrylic packing and flask closure.

Although there are continuous advancements in the field of polymers, the acrylization process and the digitization of procedures, the conventional compression molding technique is still an easier and cost-friendly substitute that is favored and practiced by a majority of laboratories. The effect observed in this in vitro study should be checked in a clinical setup for its clinical relevance. This study did not compare cross arch measurements, palatal adaptation measurements, the effect of different investing media, etc. Therefore, further studies should be planned with an increased sample size and various other parameters to check the effects of different tooth forms on denture tooth movement and their clinical relevancy. This may help in the selection of a proper processing technique for a particular tooth form, thus minimizing occlusal discrepancies and reducing occlusal corrections during laboratory and clinical remount procedures.

5. Conclusions

This study aimed to evaluate the combined effect of processing techniques (compression and injection molding) and posterior tooth forms (semi-anatomic and non-anatomic) on linear dimension changes (horizontal and vertical dimensions) in denture tooth positions following processing. Within the limitations of the study, the following conclusions can be drawn:

- In relation to dimensional changes, there was a significant interaction between the components, i.e., the posterior tooth forms and the processing technique applied.
- The injection molding technique showed smaller tooth movements in both the vertical and horizontal measurements as compared to the conventional compression molding technique.
- Non-anatomic teeth show fewer changes in tooth movement as compared to semi-anatomic teeth in both the compression and injection molding techniques.
- The changes in the position of the molar were highest in the horizontal dimension as compared to other dimensions tested for all the groups, when semi-anatomic teeth were processed using the conventional molding technique.
- The smallest differences were observed in the non-anatomic group, when the injection processing technique was used.

The clinical guideline that can be drawn from this research is that, when non-anatomic teeth are selected for complete dentures, any denture processing technique can be selected, as both cause minimal teeth movement. However, when semi-anatomic teeth are selected, an injection molding processing technique should be preferred to keep tooth movements to a minimum.

Supplementary Materials: The following supporting information can be downloaded at: https://www.mdpi.com/article/10.3390/app12147058/s1, Table S1. Shapiro Wilk Normality test used to check the normality of data without taking into account the absolute value of differences Table S2: Shapiro Wilk Normality test used to check the normality of data after absolute value of differences was taken into account.

Author Contributions: Conceptualization, M.E.S., A.P. and S.J.; methodology, M.E.S., A.P., S.J., A.H.A.H., R.A.Z., S.M.M.Z. and S.I.N.; software, A.H.A. and N.M.A.; validation, M.E.S., A.P., S.J., F.H.A. and S.S.A.; formal analysis, M.E.S., A.P., S.J., A.H.A.H., R.A.Z., S.M.M.Z. and S.I.N.; investigation, M.E.S., A.P., S.J., A.H.A.H., R.A.Z., S.M.M.Z. and S.I.N.; resources, F.H.A., A.H.A., N.M.A., H.A., S.M.A. and S.S.A.; data curation, M.E.S., A.P. and S.J.; writing—original draft preparation, M.E.S., A.P. and S.J.; writing—review and editing, M.E.S., A.P., S.J., A.H.A.H., R.A.Z., S.M.M.Z., S.I.N., F.H.A., A.H.A., N.M.A., H.A., S.M.A. and S.S.A.; visualization, F.H.A., A.H.A., N.M.A., H.A., S.M.A. and S.S.A.; supervision, M.E.S., A.P. and S.J.; project administration, M.E.S., A.P., H.A., S.M.A., A.H.A.H., R.A.Z., S.M.M.Z., S.I.N. and S.J.; funding acquisition, F.H.A., A.H.A., N.M.A., H.A., S.M.A. and S.S.A. All authors have read and agreed to the published version of the manuscript.

Funding: This research received no external funding.

Institutional Review Board Statement: The research protocol was approved by the research board at the College of Dentistry, Jazan University (Reference number: CODJU-2003I).

Informed Consent Statement: Not applicable.

Data Availability Statement: The data that support the findings of this study are available from the corresponding author upon reasonable request.

Acknowledgments: Authors would like to thank Arwa Jaber I. Mohana and Fatimah Yahya Arif for their assistance with the manuscript review process.

Conflicts of Interest: The authors declare no conflict of interest.

References

1. Petersen, P.E.; Bourgeois, D.; Bratthall, D.; Ogawa, H. Oral health information systems—Towards measuring progress in oral health promotion and disease prevention. *Bull. World Health Organ.* **2005**, *83*, 686–693. [PubMed]
2. Goldstein, G.; Kapadia, Y.; Campbell, S. Complete denture occlusion: Best evidence consensus statement. *J. Prosthodont. Off. J. Am. Coll. Prosthodont.* **2021**, *30*, 72–77. [CrossRef] [PubMed]
3. McCord, J.F.; Grant, A.A. Identification of complete denture problems: A summary. *Br. Dent. J.* **2000**, *189*, 128–134. [CrossRef] [PubMed]
4. Mainieri, E.T.; Boone, M.E.; Potter, R.H. Tooth movement and dimensional change of denture base materials using two investment methods. *J. Prosthet. Dent.* **1980**, *44*, 368–373. [CrossRef]
5. Carr, L.; Cleaton-Jones, P.; Fatti, P.; Wolfaardt, J. An experimental comparison of vertical tooth movement of 33 degrees and 0 degree teeth after denture processing procedures. *J. Oral Rehabil.* **1985**, *12*, 263–278. [CrossRef]
6. Shibayama, R.; Gennari Filho, H.; Mazaro, J.V.; Vedovatto, E.; Assuncao, W.G. Effect of flasking and polymerization techniques on tooth movement in complete denture processing. *J. Prosthodont. Off. J. Am. Coll. Prosthodont.* **2009**, *18*, 259–264. [CrossRef]
7. de Negreiros, W.A.; Consani, R.L.; Verde, M.A.; da Silva, A.M.; Pinto, L.P. The role of polymerization cycle and post-pressing time on tooth movement in complete dentures. *Braz. Oral Res.* **2009**, *23*, 467–472. [CrossRef]
8. Jamani, K.D.; Moligoda Abuzar, M.A. Effect of denture thickness on tooth movement during processing of complete dentures. *J. Oral. Rehabil.* **1998**, *25*, 725–729. [CrossRef]
9. Consani, R.L.; Domitti, S.S.; Mesquita, M.F.; Consani, S. Influence of flask closure and flask cooling methods on tooth movement in maxillary dentures. *J. Prosthodont. Off. J. Am. Coll. Prosthodont.* **2006**, *15*, 229–234. [CrossRef]
10. Abuzar, M.A.; Jamani, K.; Abuzar, M. Tooth movement during processing of complete dentures and its relation to palatal form. *J. Prosthet. Dent.* **1995**, *73*, 445–449. [CrossRef]
11. Sayed, M.E.; Swaid, S.M.; Porwal, A. Effect of cast modification on linear dimensional change of acrylic tooth position following maxillary complete denture processing. *J. Prosthodont. Off. J. Am. Coll. Prosthodont.* **2017**, *26*, 659–663. [CrossRef] [PubMed]
12. Goodacre, B.J.; Goodacre, C.J.; Baba, N.Z.; Kattadiyil, M.T. Comparison of denture tooth movement between CAD-CAM and conventional fabrication techniques. *J. Prosthet. Dent.* **2018**, *119*, 108–115. [CrossRef] [PubMed]
13. Fernandez, M.A.; Nimmo, A.; Behar-Horenstein, L.S. Digital denture fabrication in pre- and postdoctoral education: A survey of U.S. dental schools. *J. Prosthodont. Off. J. Am. Coll. Prosthodont.* **2016**, *25*, 83–90. [CrossRef] [PubMed]
14. Jackson, A.D.; Lang, B.R.; Wang, R.F. The influence of teeth on denture base processing accuracy. *Int. J. Prosthodont.* **1993**, *6*, 333–340.

15. Chintalacheruvu, V.K.; Balraj, R.U.; Putchala, L.S.; Pachalla, S. Evaluation of three different processing techniques in the fabrication of complete dentures. *J. Int. Soc. Prev. Commun. Dent.* **2017**, *7*, S18–S23. [CrossRef]
16. Nogueira, S.S.; Ogle, R.E.; Davis, E.L. Comparison of accuracy between compression- and injection-molded complete dentures. *J. Prosthet. Dent.* **1999**, *82*, 291–300. [CrossRef]
17. Yoshida, K.; Okane, H.; Nagasawa, T.; Tsuru, H. A criterion for the selection of artificial posterior teeth. *J. Oral Rehabil.* **1988**, *15*, 373–378. [CrossRef]
18. Murthy, S.S.; Murthy, G.S. Argon ion laser polymerized acrylic resin: A comparative analysis of mechanical properties of laser cured, light cured and heat cured denture base resins. *J. Int. Oral Health* **2015**, *7*, 28–34.
19. Nik, T.H.; Shahroudi, A.S.; Eraghihzadeh, Z.; Aghajani, F. Comparison of residual monomer loss from cold-cure orthodontic acrylic resins processed by different polymerization techniques. *J. Orthod.* **2014**, *41*, 30–37. [CrossRef]
20. Price, C.A. A history of dental polymers. *Aust. Prosthodont. J.* **1994**, *8*, 47–54.
21. Praveen, B.; Babaji, H.V.; Prasanna, B.G.; Rajalbandi, S.K.; Shreeharsha, T.V.; Prashant, G.M. Comparison of impact strength and fracture morphology of different heat cure denture acrylic resins: An in vitro study. *J. Int. Oral Health* **2014**, *6*, 12–16. [PubMed]
22. Becker, C.M.; Smith, D.E.; Nicholls, J.I. The comparison of denture-base processing techniques. Part I. Material characteristics. *J. Prosthet. Dent.* **1977**, *37*, 330–338. [CrossRef]
23. Consani, R.L.; Mesquita, M.F.; Sinhoreti, M.A.; Consani, S. Influence of the deflasking delay time on the displacements of maxillary denture teeth. *J. Appl. Oral Sci.* **2003**, *11*, 332–336. [CrossRef] [PubMed]
24. Wesley, R.C.; Henderson, D.; Frazier, Q.Z.; Rayson, J.H.; Ellinger, C.W.; Lutes, M.R.; Rahn, A.O.; Haley, J.V. Processing changes in complete dentures: Posterior tooth contacts and pin opening. *J. Prosthet. Dent.* **1973**, *29*, 46–54. [CrossRef]
25. Campbell, R.L. Effects of water sorption on retention of acrylic resin denture bases. *J. Am. Dent. Assoc.* **1956**, *52*, 448–454. [CrossRef]
26. Negreiros, W.A.; Consani, R.L.; Mesquita, M.F.; Sinhoreti, M.A.; Faria, I.R. Effect of flask closure method and post-pressing time on the displacement of maxillary denture teeth. *Open Dent. J.* **2009**, *3*, 21–25. [CrossRef]
27. Parvizi, A.; Lindquist, T.; Schneider, R.; Williamson, D.; Boyer, D.; Dawson, D.V. Comparison of the dimensional accuracy of injection-molded denture base materials to that of conventional pressure-pack acrylic resin. *J. Prosthodont. Off. J. Am. Coll. Prosthodont.* **2004**, *13*, 83–89. [CrossRef]
28. Sykora, O.; Sutow, E.J. Comparison of the dimensional stability of two waxes and two acrylic resin processing techniques in the production of complete dentures. *J. Oral Rehabil.* **1990**, *17*, 219–227. [CrossRef]
29. Strohaver, R.A. Comparison of changes in vertical dimension between compression and injection molded complete dentures. *J. Prosthet. Dent.* **1989**, *62*, 716–718. [CrossRef]
30. El Bahra, S.; Ludwig, K.; Samran, A.; Freitag-Wolf, S.; Kern, M. Linear and volumetric dimensional changes of injection-molded PMMA denture base resins. *Dent. Mater. Off. Publ. Acad. Dent. Mater.* **2013**, *29*, 1091–1097. [CrossRef]
31. Basso, M.F.; Nogueira, S.S.; Arioli-Filho, J.N. Comparison of the occlusal vertical dimension after processing complete dentures made with lingualized balanced occlusion and conventional balanced occlusion. *J. Prosthet. Dent.* **2006**, *96*, 200–204. [CrossRef] [PubMed]
32. Venus, H.; Boening, K.; Peroz, I. The effect of processing methods and acrylic resins on the accuracy of maxillary dentures and toothless denture bases: An in vitro study. *Quintessence Int.* **2011**, *42*, 669–677. [PubMed]
33. Peyton, F.; Craig, R. *Restorative Dental Materials*, 4th ed.; The CV Mosby Company: St. Louis, MO, USA, 1971; p. 48.
34. Zakhari, K.N. Relationship of investing medium to occlusal changes and vertical opening during denture construction. *J. Prosthet. Dent.* **1976**, *36*, 501–509. [CrossRef]
35. Shippee, R.W. Control of increased vertical dimension of compression-molded dentures. *J. Prosthet. Dent.* **1961**, *11*, 1080–1085. [CrossRef]
36. Atkinson, H.; Grant, A.J.A.D.J. An investigation into tooth movement during the packing and polymerizing of acrylic resin denture base materials. *Austr. Dent. J.* **1962**, *7*, 101–108. [CrossRef]
37. Atashrazm, P.; Alavijeh, L.Z.; Afshar, M.S. Influence of the fast-processing technique on the number of the occlusal contacts and occlusal vertical dimension of complete dentures. *J. Contemp. Dent. Pract.* **2011**, *12*, 84–90. [CrossRef] [PubMed]
38. Anderson, G.C.; Schulte, J.K.; Arnold, T.G. Dimensional stability of injection and conventional processing of denture base acrylic resin. *J. Prosthet. Dent.* **1988**, *60*, 394–398. [CrossRef]
39. Chuchulska, B.; Yankov, S.; Todorov, R.; Ivanova, D.; Kalachev, Y. Injection Shrinkage and Water Sorption of Some Thermoplastic Dental Materials. *Pesqui. Bras. Odontopediatr. Clín. Integr.* **2019**, *19*, e4474. [CrossRef]
40. Woelfel, J.B.; Paffenbarger, G.C.; Sweeney, W.T. Dimensional changes occurring in dentures during processing. *J. Am. Dent. Assoc.* **1960**, *61*, 413–430. [CrossRef]

applied sciences

MDPI

Article

Tooth Shade Relationship with Age, Gender, and Skin Color in a Saudi Population: A Cross-Sectional Study

Hussain Alsayed [1,*], Razan Alaqeely [2], Nawaf Almazrouei [3], Mohammed Alzahrani [3], Faris Alzahrani [3], Abdulaziz Bin Oun [3] and Abdulmonem Alshihri [1]

1 Department of Prosthetic Dental Sciences, College of Dentistry, King Saud University, Riyadh 11545, Saudi Arabia; aalshihre@ksu.edu.sa

2 Department of Periodontics and Community Dentistry, College of Dentistry, King Saud University, Riyadh 11545, Saudi Arabia; raqeely@ksu.edu.sa

3 Ministry of Health, Riyadh 11992, Saudi Arabia; nawafalmazrouei@gmail.com (N.A.); lmadol@hotmail.com (M.A.); faris.saz95@gmail.com (F.A.); abinoun@gmail.com (A.B.O.)

* Correspondence: halsayed@ksu.edu.sa

Abstract: ***Objectives:*** The aim of this study was to assess the relationship between tooth shade among different groups of patients according to their age, gender, and skin color in a Saudi population. ***Materials and methods:*** Participants were divided based on age into Group 1 (10–20 years), Group 2 (21–30 years), Group 3 (31–40 years), and Group 4 (41+ years), and according to gender. Tooth shade was measured by Vita easyshade, Shade scanner, 3D Master shade system. The skin color was determined according to the Firzpatrick Scale. It consists of six shades, namely: I, II, III, IV, V, and VI. The skin complexion of the participants was divided into six categories: white/very fair, fair, light brown, moderate brown, dark brown, and black. ***Results:*** One hundred and ninety-eight individuals were recruited. Around 70% were males. Females had 25.4% A2 followed by 22% A1, and 22% A3 shade types, while males had B3 shade (18%) followed by A2 and A3 (15.8%). A statistically significant difference was observed between shade and gender ($p < 0.05$). A statistically significant difference was observed between shade and age group ($p < 0.05$), where increased age was correlated with darker teeth shades. Shade A1 was correlated with type I skin color in 57.1% of individuals. Skin color type II had A2 as a dominant shade by 34.1%. A2 and A3 shades were equally observed in skin color III by 20.3%. Overall, statistically significant differences were observed between shade and skin color groups ($p < 0.05$). ***Conclusions:*** In conclusion, the most frequent classical shade noted among male and female participants was shade type A, which represents reddish brownish. Group 2 (21–30 years) had the B3 shade as the most prominent shade type among age groups. Gender, age, and skin types all showed a significant relation with the tooth shade.

Keywords: tooth shade; skin color; shade and age; skin and shade; Saudi

Citation: Alsayed, H.; Alaqeely, R.; Almazrouei, N.; Alzahrani, M.; Alzahrani, F.; Bin Oun, A.; Alshihri, A. Tooth Shade Relationship with Age, Gender, and Skin Color in a Saudi Population: A Cross-Sectional Study. *Appl. Sci.* **2022**, *12*, 6315. https://doi.org/10.3390/app12136315

Academic Editor: Mary Anne Melo

Received: 11 May 2022
Accepted: 18 June 2022
Published: 21 June 2022

1. Introduction

Esthetic dentistry has been a concern for patients visiting dental clinics. Patients have more expectations and are looking for higher esthetic results as a consequence of increased awareness and the impact of social media [1]. Usually, patients will assess the dentist's work by function, comfort, speech, and esthetics. The form and alignment of the teeth determine the beauty of the smile; the balance between the color of the teeth and the soft tissues is considered fundamental in determining the satisfaction of the person's dental appearance [2,3]. Systems evolved in defining the shade of teeth depend on three attributes: Hue, Chroma, and Value. The Hue is described as the character of appearance, which is discernible as (red, green, blue, etc.) and depends on the wavelength of each color. Chroma is the saturation of the color as it distinguishes a strong color from a weak one. The value is the color brightness, which compiles the lightness or darkness of the color [4].

One of the obstacles that most dentists are facing is matching a restoration to a natural tooth shade. However, patients described pearly white teeth as preferable [2,5]. Identifying the tooth shade will aid dentists in using fewer amounts of shade tabs and guides required to suit the patients' desires. This will also aid in providing better esthetic results according to the patients' age, gender, and skin color. According to Jahaniri et al., young patients were more likely to have teeth with a high value (lighter teeth) [6,7]. Factors like gender, age, and skin color may contribute to the method of shade selection [8,9]. In Sudan, for instance, shades A3, A2, and A1 were the most common tooth shades, respectively [10]. Karaman et al., in Turkey, reported that shade A2 was most observed in central and lateral incisors for all age groups. For the canines, B3 was most observed for the youngest two groups and A3.5 for the oldest two groups [11]. A study conducted in Jordan comparing tooth shade to skin color revealed that Jordanians have a light tooth value and that people with a dark skin complex have lighter teeth [12]. The skin color of an individual might help the prosthodontist in picking the artificial teeth shade in case of a complete denture [8]. Jahangiri et al. concluded that 50% of persons with low-value teeth had fair skin, whereas 17% had dark skin [6]. However, a study in Nepal [13], and another in India [14], exhibited no difference between all skin colors and tooth shades. Another study conducted in Korea reported that lighter and less chromatic central incisors were characteristics of female subjects in all age groups [15]. Gómez-Polo et al. revealed that age was stronger than gender in all color coordinates. However, females have lighter teeth than males in the plots [9]. Karaman et al. also showed that gender is statistically not related to a color value or Chroma [11]. In Saudi Arabia, a study in the southern region found that the shade gets darker and more yellow with age [16]. However, one ethnic group was examined in this study and could not be generalized to a Saudi Arabia population. Nowadays, high expectations from social media and patient awareness demand a quality level of training and knowledge regarding all aspects of esthetic dentistry. The process of shade selection is challenging because it is a subjective process. Therefore, patient-centered treatments and guidance in shade selection lead to a successful outcome when executed with prior knowledge of how the color is perceived.

Therefore, this research aims to assess teeth shades and their relationship with age, gender, and skin color in a Saudi population.

2. Materials and Methods

The approval of the Institutional Review Board was granted to conduct a cross-sectional study on diverse age groups. The sample size was calculated using a G power software to be 196 Saudi participants (n = 139 males, n = 59 females) visiting the Dental University Hospital (DUH) at King Saud University in Riyadh, Saudi Arabia. Included participants were only Saudi to limit the heterogenicity of the skin color and to homogenize the skin color in relation to the age. The inclusion criteria were healthy participants, above the age of 15 years, with fully erupted maxillary anterior teeth. Participants with no history of periodontal disease, bleaching, or active orthodontic treatment were included. The exclusion criteria were participants with anterior teeth having intrinsic and extrinsic stains, smoking, tobacco chewing, or developmental defects. Furthermore, teeth that are endodontically treated, restored, or carious were excluded. Any participant who reported tanned skin, dermatological disease, or undergoing any dermatological treatment was also excluded. Participants were divided based on gender (male and female) as well as age (Group 1: 15–20 years, Group 2: 21–30 years, Group 3: 31–40 years, and Group 4: 41 years and more).

Tooth shade was obtained by VITA Easy shade Advance 4.0® spectrophotometer (VITA Zahnfabrik, Bad Sackingen, Germany), which was calibrated following the manufacturer's instructions. The recorded shade was the closest to the classical shade guide. Prophylaxis was not carried out before shade selection because dehydration affects the tooth shade and requires hours to regain the original shade [17,18]. Instead, they were asked to brush for 3 min prior. Then, the teeth were wiped with a sterile gauze [15]. The tip of the spectrophotometer was held at a 90° angle against the middle aspect of the labial surface of tooth #11.

The skin phototypes of all participants were selected utilizing the Fitzpatrick skin phototypes criteria, skin phototype I: pale white skin, skin phototype II: fair skin, skin phototype III: darker white skin, skin phototype IV: light brown skin, skin phototype V: brown skin, and phototype VI: dark brown or black skin [19]. The skin color evaluation method was determined following Treesirichod's protocol: a room with fluorescent lighting, with no interfering outdoor sunlight, and at 20–30 min after enrollment [20]. The sample size was calculated using the G power software considering a power of 0.95, the effect size of 0.23, and the alpha set at 0.05 to be 196 participants. Data were analyzed by statistical software (SPSS 22.0, SPSS) using descriptive statistics, Pearson product-moment correlation, and chi-square tests. The statistical significance level was set at $p < 0.05$.

3. Result

One hundred and ninety-eight individuals were recruited from December 2020 until February 2021. Around 70% were males (n = 139) and the rest were females (n = 59). Participants were divided into five groups based on age where the majority (n = 78) of individuals were from Group 2 (21–30 years), followed by 45 individuals from Group 3 (31–40 years), 36 individuals from Group 1 (15–20 years), 25 individuals from Group 4 (41–50 years), and 14 individuals from Group 5 (51 years and more).

In the current study, the most frequent shade noted among male and female participants (n = 120) was shade type A, which represents reddish brownish. The second most frequent shade was type B (n = 49), followed by 20 participants with shade type C, and nine with shade D (Table 1). Most female participants had an A2 shade representing 25.4%, followed by A2 with 22%, and A3 with another 22% individually, while males had a B3 shade (18%) followed by A2 and A3 (15.8%) (Figure 1). Overall, the results showed a significant relationship between tooth shade and participant's gender, with a p-value of less than 0.05 (Table 2).

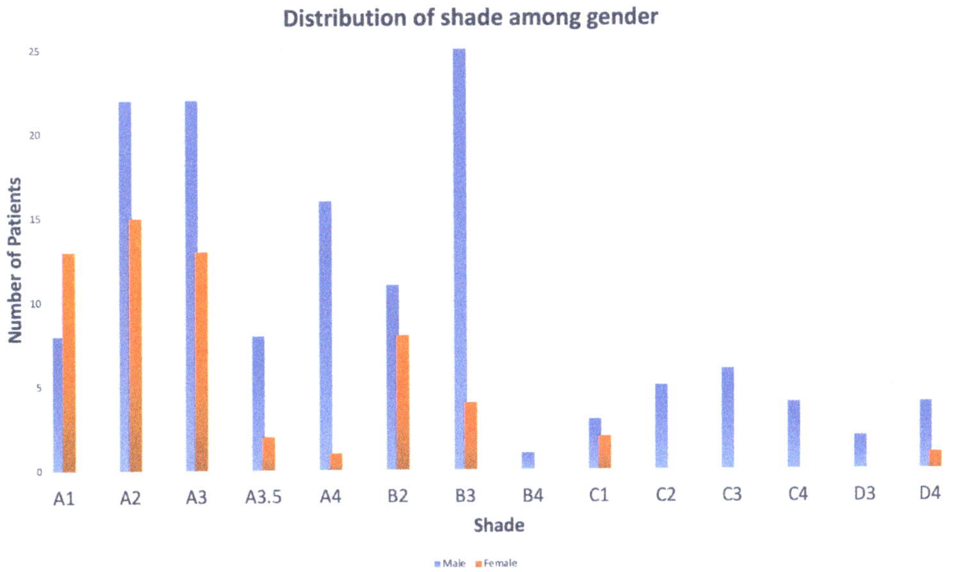

Figure 1. Relationship between teeth shade and among male and female participants.

Table 1. Relationship between teeth shade and gender. Most participants had an A2 shade.

Relationship between Shade and Gender *				
Shade		**Gender**		**Total**
		Male	**Female**	
A1	No. Participants	8	13	21
	% within Shade	38.1%	61.9%	100.0%
	% within Gender	5.8%	22.0%	10.6%
A2	No. Participants	22	15	37
	% within Shade	59.5%	40.5%	100.0%
	% within Gender	15.8%	25.4%	18.7%
A3	No. Participants	22	13	35
	% within Shade	62.9%	37.1%	100.0%
	% within Gender	15.8%	22.0%	17.7%
A3.5	No. Participants	8	2	10
	% within Shade	80.0%	20.0%	100.0%
	% within Gender	5.8%	3.4%	5.1%
A4	No. Participants	16	1	17
	% within Shade	94.1%	5.9%	100.0%
	% within Gender	11.5%	1.7%	8.6%
B2	No. Participants	11	8	19
	% within Shade	57.9%	42.1%	100.0%
	% within Gender	7.9%	13.6%	9.6%
B3	No. Participants	25	4	29
	% within Shade	86.2%	13.8%	100.0%
	% within Gender	18.0%	6.8%	14.6%
B4	No. Participants	1	0	1
	% within Shade	100.0%	0.0%	100.0%
	% within Gender	0.7%	0.0%	0.5%
C1	No. Participants	3	2	5
	% within Shade	60.0%	40.0%	100.0%
	% within Gender	2.2%	3.4%	2.5%
C2	No. Participants	5	0	5
	% within Shade	100.0%	0.0%	100.0%
	% within Gender	3.6%	0.0%	2.5%
C3	No. Participants	6	0	6
	% within Shade	100.0%	0.0%	100.0%
	% within Gender	4.3%	0.0%	3.0%
C4	No. Participants	4	0	4
	% within Shade	100.0%	0.0%	100.0%
	% within Gender	2.9%	0.0%	2.0%
D3	No. Participants	2	0	2
	% within Shade	100.0%	0.0%	100.0%
	% within Gender	1.4%	0.0%	1.0%
D4	No. Participants	4	1	5
	% within Shade	80.0%	20.0%	100.0%
	% within Gender	2.9%	1.7%	2.5%
Total	No. Participants	139	59	198
	% within Shade	70.2%	29.8%	100.0%
	% within Gender	100.0%	100.0%	100.0%

* $p < 0.05$.

Table 2. Pearson Chi-Square test shows the significant level between shade and gender.

Chi-Square Tests			
	Value	df	Asymptotic Significance (2-Sided)
Pearson Chi-Square	32.294 [a]	15	0.006
Likelihood Ratio	38.688	15	0.001
Linear-by-Linear Association	16.805	1	0.000
N of Valid Cases	198		

[a] 19 cells (59.4%) have expected count less than 5. The minimum expected count is 0.30.

According to the age, 33.3% of Group 1 (10–20 years) had shade A2 while Group 2 (21–30 years) and 3 (31–40 years) had 17.9% and 22.2% of shade B3, respectively. Shade A3 accounted for 20% of individuals in Group 4 and 28.6% of Group 5 individuals (Table 3) (Figure 2). A statistically significant difference was observed between shade and age groups ($p < 0.05$) (Table 4). Skin color type IV appeared to be the most dominant type with a total of 67 individuals, followed by skin color type III with a total of 64 individuals. Forty-one individuals had type II skin color, 17 individuals had type V skin color, then skin type I and VI with 7 and 2 individuals, respectively (Table 5).

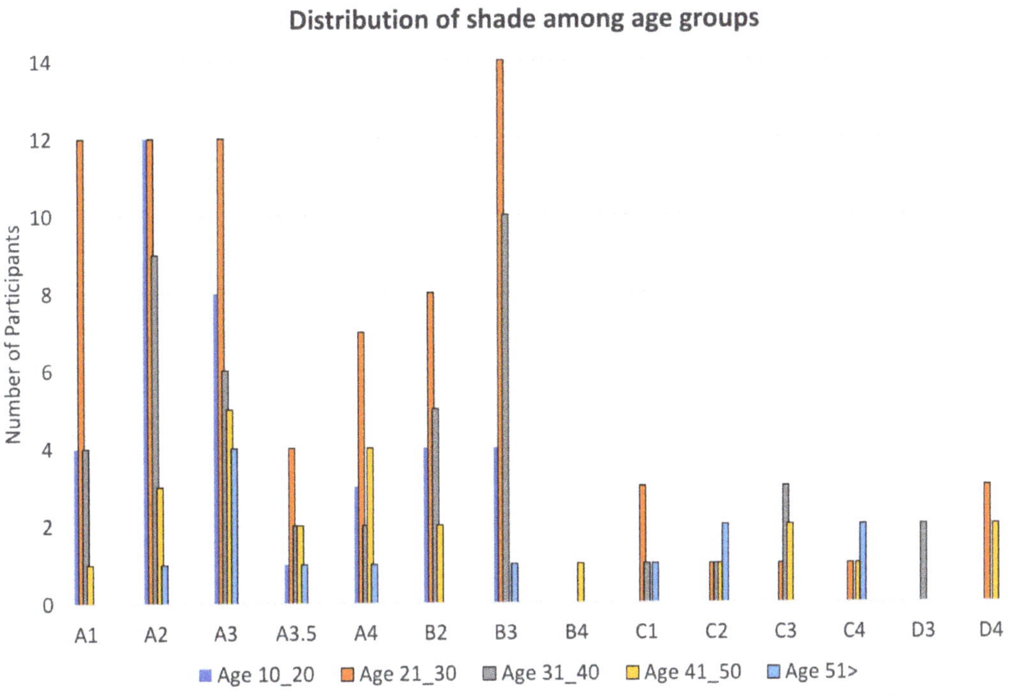

Figure 2. Relationship between teeth shade and age groups. A2 was the most frequent shade in all age groups.

Table 3. Relationship between teeth shade and age groups showed a statistical significance.

Relationship between Shade and Age *							
Shade		**Age**					**Total**
		10–20	**21–30**	**31–40**	**41–50**	**51>**	
A1	No. Participants	4	12	4	1	0	21
	% within Shade	19.0%	57.1%	19.0%	4.8%	0.0%	100.0%
	% within Age	11.1%	15.4%	8.9%	4.0%	0.0%	10.6%
A2	No. Participants	12	12	9	3	1	37
	% within Shade	32.4%	32.4%	24.3%	8.1%	2.7%	100.0%
	% within Age	33.3%	15.4%	20.0%	12.0%	7.1%	18.7%
A3	No. Participants	8	12	6	5	4	35
	% within Shade	22.9%	34.3%	17.1%	14.3%	11.4%	100.0%
	% within Age	22.2%	15.4%	13.3%	20.0%	28.6%	17.7%
A3.5	No. Participants	1	4	2	2	1	10
	% within Shade	10.0%	40.0%	20.0%	20.0%	10.0%	100.0%
	% within Age	2.8%	5.1%	4.4%	8.0%	7.1%	5.1%
A4	No. Participants	3	7	2	4	1	17
	% within Shade	17.6%	41.2%	11.8%	23.5%	5.9%	100.0%
	% within Age	8.3%	9.0%	4.4%	16.0%	7.1%	8.6%
B2	No. Participants	4	8	5	2	0	19
	% within Shade	21.1%	42.1%	26.3%	10.5%	0.0%	100.0%
	% within Age	11.1%	10.3%	11.1%	8.0%	0.0%	9.6%
B3	No. Participants	4	14	10	0	1	29
	% within Shade	13.8%	48.3%	34.5%	0.0%	3.4%	100.0%
	% within Age	11.1%	17.9%	22.2%	0.0%	7.1%	14.6%
B4	No. Participants	0	0	0	1	0	1
	% within Shade	0.0%	0.0%	0.0%	100.0%	0.0%	100.0%
	% within Age	0.0%	0.0%	0.0%	4.0%	0.0%	0.5%
C1	No. Participants	0	3	1	0	1	5
	% within Shade	0.0%	60.0%	20.0%	0.0%	20.0%	100.0%
	% within Age	0.0%	3.8%	2.2%	0.0%	7.1%	2.5%
C2	No. Participants	0	1	1	1	2	5
	% within Shade	0.0%	20.0%	20.0%	20.0%	40.0%	100.0%
	% within Age	0.0%	1.3%	2.2%	4.0%	14.3%	2.5%
C3	No. Participants	0	1	3	2	0	6
	% within Shade	0.0%	16.7%	50.0%	33.3%	0.0%	100.0%
	% within Age	0.0%	1.3%	6.7%	8.0%	0.0%	3.0%
C4	No. Participants	0	1	0	1	2	4
	% within Shade	0.0%	25.0%	0.0%	25.0%	50.0%	100.0%
	% within Age	0.0%	1.3%	0.0%	4.0%	14.3%	2.0%
D3	No. Participants	0	0	2	0	0	2
	% within Shade	0.0%	0.0%	100.0%	0.0%	0.0%	100.0%
	% within Age	0.0%	0.0%	4.4%	0.0%	0.0%	1.0%
D4	No. Participants	0	3	0	2	0	5
	% within Shade	0.0%	60.0%	0.0%	40.0%	0.0%	100.0%
	% within Age	0.0%	3.8%	0.0%	8.0%	0.0%	2.5%
Total	No. Participants	36	78	45	25	14	198
	% within Shade	18.2%	39.4%	22.7%	12.6%	7.1%	100.0%
	% within Age	100.0%	100.0%	100.0%	100.0%	100.0%	100.0%

* $p < 0.05$.

Table 4. Pearson Chi-Square test shows the significant level between shade and age groups.

Chi-Square Tests			
	Value	**df**	**Asymptotic Significance (2-Sided)**
Pearson Chi-Square	96.008 [a]	60	0.002
Likelihood Ratio	82.950	60	0.027
Linear-by-Linear Association	16.142	1	0.000
N of Valid Cases	198		

[a] 68 cells (85.0%) have expected count less than 5. The minimum expected count is 0.07.

Table 5. Relationship between shade and skin types. Skin color type IV appeared to be the most dominant type.

Relationship between Shade and Skin Color *								
Shade		Skin Color						Total
		I	II	III	IV	V	VI	
A1	No. Participants	4	5	8	1	3	0	21
	% within Shade	19.0%	23.8%	38.1%	4.8%	14.3%	0.0%	100.0%
	% within Skin color	57.1%	12.2%	12.5%	1.5%	17.6%	0.0%	10.6%
A2	No. Participants	1	14	13	5	3	1	37
	% within Shade	2.7%	37.8%	35.1%	13.5%	8.1%	2.7%	100.0%
	% within Skin color	14.3%	34.1%	20.3%	7.5%	17.6%	50.0%	18.7%
A3	No. Participants	1	2	13	18	1	0	35
	% within Shade	2.9%	5.7%	37.1%	51.4%	2.9%	0.0%	100.0%
	% within Skin color	14.3%	4.9%	20.3%	26.9%	5.9%	0.0%	17.7%
A3.5	No. Participants	0	2	0	6	2	0	10
	% within Shade	0.0%	20.0%	0.0%	60.0%	20.0%	0.0%	100.0%
	% within Skin color	0.0%	4.9%	0.0%	9.0%	11.8%	0.0%	5.1%
A4	No. Participants	1	3	5	6	2	0	17
	% within Shade	5.9%	17.6%	29.4%	35.3%	11.8%	0.0%	100.0%
	% within Skin color	14.3%	7.3%	7.8%	9.0%	11.8%	0.0%	8.6%
B2	No. Participants	0	6	10	2	0	1	19
	% within Shade	0.0%	31.6%	52.6%	10.5%	0.0%	5.3%	100.0%
	% within Skin color	0.0%	14.6%	15.6%	3.0%	0.0%	50.0%	9.6%
B3	No. Participants	0	4	6	17	2	0	29
	% within Shade	0.0%	13.8%	20.7%	58.6%	6.9%	0.0%	100.0%
	% within Skin color	0.0%	9.8%	9.4%	25.4%	11.8%	0.0%	14.6%
B4	No. Participants	0	0	0	1	0	0	1
	% within Shade	0.0%	0.0%	0.0%	100.0%	0.0%	0.0%	100.0%
	% within Skin color	0.0%	0.0%	0.0%	1.5%	0.0%	0.0%	0.5%
C1	No. Participants	0	1	1	3	0	0	5
	% within Shade	0.0%	20.0%	20.0%	60.0%	0.0%	0.0%	100.0%
	% within Skin color	0.0%	2.4%	1.6%	4.5%	0.0%	0.0%	2.5%
C2	No. Participants	0	2	2	1	0	0	5
	% within Shade	0.0%	40.0%	40.0%	20.0%	0.0%	0.0%	100.0%
	% within Skin color	0.0%	4.9%	3.1%	1.5%	0.0%	0.0%	2.5%
C3	No. Participants	0	0	3	2	1	0	6
	% within Shade	0.0%	0.0%	50.0%	33.3%	16.7%	0.0%	100.0%
	% within Skin color	0.0%	0.0%	4.7%	3.0%	5.9%	0.0%	3.0%
C4	No. Participants	0	0	2	2	0	0	4
	% within Shade	0.0%	0.0%	50.0%	50.0%	0.0%	0.0%	100.0%
	% within Skin color	0.0%	0.0%	3.1%	3.0%	0.0%	0.0%	2.0%
D3	No. Participants	0	1	0	0	1	0	2
	% within Shade	0.0%	50.0%	0.0%	0.0%	50.0%	0.0%	100.0%
	% within Skin color	0.0%	2.4%	0.0%	0.0%	5.9%	0.0%	1.0%
D4	No. Participants	0	1	1	1	2	0	5
	% within Shade	0.0%	20.0%	20.0%	20.0%	40.0%	0.0%	100.0%
	% within Skin color	0.0%	2.4%	1.6%	1.5%	11.8%	0.0%	2.5%
Total	No. Participants	7	41	64	67	17	2	198
	% within Shade	3.5%	20.7%	32.3%	33.8%	8.6%	1.0%	100.0%
	% within Skin color	100.0%	100.0%	100.0%	100.0%	100.0%	100.0%	100.0%

* $p < 0.05$.

Shade A1 was related to type I skin color in 57.1% of individuals. Skin color type II had A2 as a dominant shade by 34.1%. A2 and A3 shades were equally observed in skin color type III by 20.3%. A3 shade was observed in 26.9% with skin color IV, while A1 and A2 shared the most observed shade in skin color V by 17.6% (Table 5). The relationship between tooth shade and skin color is exhibited in (Figure 3). The results showed statistically significant differences between shade and skin color groups ($p < 0.05$) (Table 6).

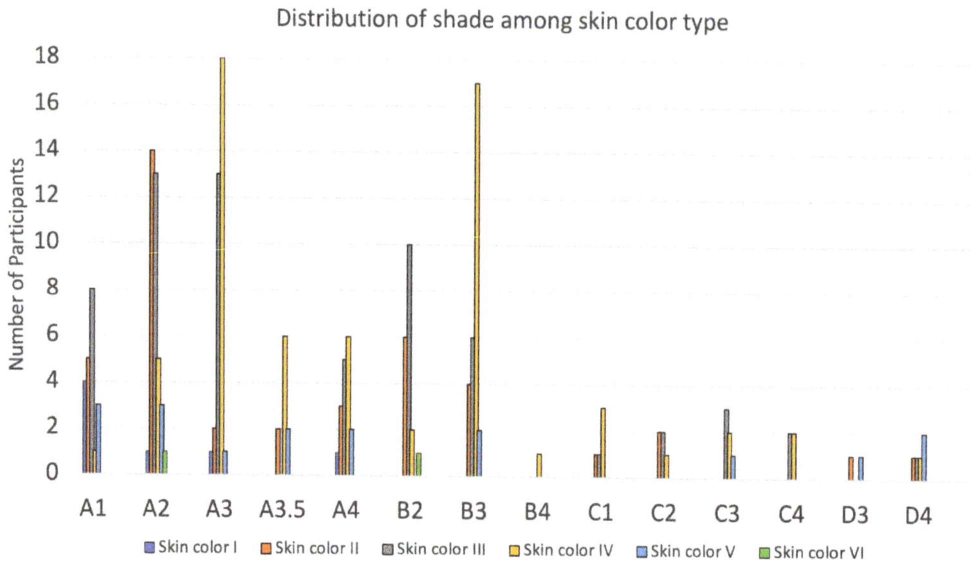

Figure 3. Relationship between teeth shade and skin color. Skin color type II had A2 as a dominant shade type.

Table 6. Pearson Chi-Square test shows the significant level between shade and skin type.

	Chi-Square Tests		
	Value	**df**	**Asymptotic Significance (2-Sided)**
Pearson Chi-Square	96.689 [a]	75	0.047
Likelihood Ratio	100.032	75	0.028
Linear-by-Linear Association	7.172	1	0.007
N of Valid Cases	198		

[a] 81 cells (84.4%) have expected count less than 5. The minimum expected count is 0.01.

4. Discussion

Selecting a proper tooth shade is considered a complex process during prosthetic rehabilitation, which requires fundamental knowledge of color and esthetic. The dentist's skills in determining the right shade play a role in the success of treatment and patient satisfaction. Multiple factors such as light and background affect tooth color [21]. Some studies prefer using conventional shade tabs, while others consider digital devices for more accuracy and precision [22,23]. Digital devices can detect more data about tooth shade, such as lightness, Chroma, and Hue, which can aid in mimicking the color of adjacent natural teeth.

According to the present study, there was a relation between shade, gender, and skin type. The participants of this cross-sectional study belong to a specific ethnic group (Middle eastern) in a specific country (Saudi Arabia). Considering the regional limitation of the data collection and sample size, the most dominant tooth shade observed was shape A2. Labban et al. studied the Saudi population's perception of their preference for the desired tooth shade. In the above-mentioned study, the researchers provided the participants with a questionnaire consisting of images that had been modified digitally to illustrate different skin and tooth shade combinations. It was found that people with lighter skin preferred to have lighter teeth shade [9]. On the contrary, data on the participants' skin type and tooth shade were obtained directly without involving participants' subjectivity in the data acquisition. Skin color might be valuable when selecting teeth shade for edentulous patients. In this study, an observation of shade relation to skin color groups was statistically significant. People with type I skin tones possess lighter tooth shade (A1 type). Furthermore, darker skin tones such as type IV and V possess darker tooth shade (A3, A3.5 types). This finding contradicts the previous work of Jahangiri et al. and Al-Nsour et al., who reported that individuals with dark skin color have lighter teeth in comparison to individuals with light skin color [6,12]. This might contribute to different ethnic populations and regional limitations.

The hypothesis that males tend to have darker shades than females was reported in previous studies. Gómez-Polo et al. study showed that 25.4% of females had A2 shade while 18% of males had B3 Shade [9]. Kim et al. reported that females' teeth are lighter and less chromatic than males' teeth [15]. In the current study, females tend to have a lighter shade and less Chroma than males, which was in agreement with previous studies.

Based on our study, aging affects tooth color. Older participants tend to have darker teeth than younger individuals. Haralur observed that teeth become darker with age due to multiple factors such as enamel thickness reduction due to wear and secondary dentin deposition [16]. Furthermore, Karaman et al. reported that a statistically significant difference between age groups was valid for central and lateral incisors [24].

5. Conclusions

Within the limitations of this study, it can be assumed that females tend to have a lighter teeth shade than males. Younger individuals have lighter teeth shades than older people in the Saudi population. The process of shade selection should reflect on the relation between skin color and teeth shade. In terms of skin type, type IV was the most common among the Saudi population. Moreover, people with type I skin tones possess lighter teeth.

Author Contributions: Conceptualization and manuscript write-up, H.A. and R.A.; Formal analysis, A.A.; Investigation, N.A.; Methodology, M.A., F.A. and A.B.O. All authors have read and agreed to the published version of the manuscript.

Funding: This research received no external funding.

Institutional Review Board Statement: The study was conducted in accordance approved by the Institutional Review Board at King Saud University (Ref. No. 21/0010/IRB on 4 January 2021).

Informed Consent Statement: Informed consent was obtained from all subjects involved in the study.

Data Availability Statement: The data that support the findings of this study are available from the corresponding author upon reasonable request.

Conflicts of Interest: The authors declare no conflict of interest.

References

1. Mayekar, S. Shades of a color. Illusion or reality? *Dent. Clin. N. Am.* **2001**, *45*, 155–172, vii. [CrossRef]
2. Qualtrough, A.J.; Burke, F.J. A look at dental esthetics. *Quintessence Int.* **1994**, *25*, 7–14. [PubMed]
3. Tam, W.K.; Lee, H.J. Dental shade matching using a digital camera. *J. Dent.* **2012**, *40*, 3–10. [CrossRef] [PubMed]
4. Fujimoto, J.; Land, M.F.; Rosenstiel, S.F. *Contemporary Fixed Prosthodontics*; Mosby Elsevier: St. Louis, MI, USA, 2015.
5. Alkhatib, M.N.; Holt, R.; Bedi, R. Prevalence of self-assessed tooth discolouration in the United Kingdom. *J. Dent.* **2004**, *32*, 561–566. [CrossRef]
6. Jahangiri, L.; Reinhardt, S.B.; Mehra, R.V.; Matheson, P.B. Relationship Between Tooth Shade Value and Skin Color: An Observational Study. *J. Prosthet. Dent.* **2002**, *87*, 149–152. [CrossRef]

7. Naveh, G.; Grossman, Y.; Bachner, Y.G.; Levin, L. Patients' self-perception of tooth shade in relation to professionally objective evaluation. *Quintessence Int.* **2010**, *41*, e80–e83.
8. Haralur, S.B.; Dibas, A.M.; Almelhi, N.A.; Al-Qahtani, D.A. The Tooth and Skin Colour Interrelationship across the Different Ethnic Groups. *Int. J. Dent.* **2014**, *2014*, 146028. [CrossRef]
9. Gómez-Polo, C.; Montero, J.; Gómez-Polo, M.; de Parga, J.A.M.V. Celemin-Viñuela A. Natural Tooth Color Estimation Based on Age and Gender. *J. Prosthodont.* **2017**, *26*, 107–114. [CrossRef]
10. Elamin, H.O.; Abubakr, N.H.; Ibrahim, Y.E. Identifying the tooth shade in group of patients using Vita Easyshade. *Eur. J. Dent.* **2015**, *9*, 213–217. [CrossRef]
11. Karaman, T.; Altintas, E.; Eser, B.; Yildirim, T.T.; Oztekin, F.; Bozoglan, A. Spectrophotometric Evaluation of Anterior Maxillary Tooth Color Distribution According to Age and Gender. *J. Prosthodont.* **2019**, *28*, e96–e102. [CrossRef]
12. Al-Nsour, H.F.; Al-Zoubi, T.T.; Al-Rimawi, A.S. Relationship between tooth value and skin color in patients visiting Royal Medical Services clinics of Jordan. *Electron. Physician* **2018**, *10*, 6448–6453. [CrossRef] [PubMed]
13. Pradhan, D.; Shrestha, L.; Lohani, J. Tooth Shade and Skin Colour: A Descriptive Cross-Sectional Study. *J. Nepal Med. Assoc.* **2020**, *58*, 144–147. [CrossRef]
14. Veeraganta, S.K.; Savadi, R.C.; Baroudi, K.; Nassani, M.Z. Differences in tooth shade value according to age, gender and skin color: A pilot study. *J. Indian Prosthodont. Soc.* **2015**, *15*, 138–141. [CrossRef] [PubMed]
15. Kim, H.K. A study on the color distribution of natural teeth by age and gender in the Korean population with an intraoral spectrophotometer. *J. Esthet. Restor. Dent.* **2018**, *30*, 408–414. [CrossRef]
16. Brodbelt, R.H.; O'Brien, W.J.; Fan, P.L.; Frazer-Dib, J.G.; Yu, R. Translucency of human dental enamel. *J. Dent. Res.* **1981**, *60*, 1749–1753. [CrossRef]
17. Miller, A.; Long, J.; Cole, J.; Staffanou, R. Shade selection and laboratory communication. *Quintessence Int.* **1993**, *24*, 305–309.
18. Sachdeva, S. Fitzpatrick skin typing: Applications in dermatology. *Indian J. Dermatol. Venereol. Leprol.* **2009**, *75*, 93–96. [CrossRef]
19. Treesirichod, A.; Chansakulporn, S.; Wattanapan, P. Correlation between skin color evaluation by skin color scale chart and narrowband reflectance spectrophotometer. *Indian J. Dermatol.* **2014**, *59*, 339–342. [CrossRef]
20. Ardu, S.; Braut, V.; Di Bella, E.; Lefever, D. Influence of background on natural tooth colour coordinates: An in vivo evaluation. *Odontology* **2014**, *102*, 267–271. [CrossRef]
21. Dozić, A.; Kleverlaan, C.J.; El-Zohairy, A.; Feilzer, A.J.; Khashayar, G. Performance of five commercially available tooth color-measuring devices. *J. Prosthodont.* **2007**, *16*, 93–100. [CrossRef]
22. Chen, H.; Huang, J.; Dong, X.; Qian, J.; He, J.; Qu, X.; Lu, E. A systematic review of visual and instrumental measurements for tooth shade matching. *Quintessence Int.* **2012**, *43*, 649–659. [PubMed]
23. Labban, N.; Al-Otaibi, H.; Alayed, A.; Alshankiti, K.; Al-Enizy, M.A. Assessment of the influence of gender and skin color on the preference of tooth shade in Saudi population. *Saudi Dent. J.* **2017**, *29*, 102–110. [CrossRef] [PubMed]
24. Haralur, S.B. Effect of Age on Tooth Shade, Skin Color and Skin-Tooth Color Interrelationship in Saudi Arabian Subpopulation. *J. Int. Oral Health* **2015**, *7*, 33–36. [PubMed]

applied sciences

MDPI

Article

Bacterial Disinfection of Polymethyl Methacrylate (PMMA) Resin Polymer Using Low Level Microwave Irradiation and Denture Cleaning Agent

Ziaullah Choudhry [1,*], Syed M. R. Kazmi [2], Sofia Malik [3], Zulfiqar A. Mirani [4], Shahbaz Ahmed [5], Fazal Ur Rehman Qazi [5], Muhammad Adeel Ahmed [6], Mohammed Alrabiah [7], Khulud A. AlAali [8], Fahim Vohra [7,*] and Tariq Abduljabbar [7]

1 Prosthodontics Department, Dr. Ishrat-Ul-Ebad Khan Institute of Oral Health Sciences, Dow University of Health Sciences, Karachi 74200, Pakistan
2 Department of Surgery, Section of Dentistry, Aga Khan University, Karachi 74800, Pakistan; syed.murtaza@aku.edu
3 Department of Science of Dental Material, Dr. Ishrat-Ul-Ebad Khan Institute of Oral Health Sciences, Dow University of Health Sciences, Karachi 74200, Pakistan; sofia.malik@duhs.edu.pk
4 Microbiology Section, Pakistan Council for Scientific and Industrial Research (PCSIR), Laboratories Complex, Karachi 75270, Pakistan; mirani_mrsa@yahoo.com
5 Operative Dentistry Department, Dr. Ishrat-Ul-Ebad Khan Institute of Oral Health Sciences, Dow University of Health Sciences, Karachi 74200, Pakistan; saj119@hotmail.com (S.A.); qazi.rehman@duhs.edu.pk (F.U.R.Q.)
6 Department of Restorative Dental Sciences, College of Dentistry, King Faisal University, Al Ahsa 31982, Saudi Arabia; mshakeel@kfu.edu.sa
7 Prosthetic Dental Science Department, College of Dentistry, King Saud University, Riyadh 11451, Saudi Arabia; mhalrabiah@ksu.edu.sa (M.A.); tajabbar@ksu.edu.sa (T.A.)
8 Department of Clinical Dental Sciences, College of Dentistry, Princess Nourah Bint Abdulrahman University, Riyadh 11564, Saudi Arabia; kaalaali@pnu.edu.sa
* Correspondence: ziaullah.choudhry@duhs.edu.pk (Z.C.); fvohra@ksu.edu.sa (F.V.); Fax: +966-14678639 (F.V.)

Citation: Choudhry, Z.; Kazmi, S.M.R.; Malik, S.; Mirani, Z.A.; Ahmed, S.; Rehman Qazi, F.U.; Ahmed, M.A.; Alrabiah, M.; AlAali, K.A.; Vohra, F.; et al. Bacterial Disinfection of Polymethyl Methacrylate (PMMA) Resin Polymer Using Low Level Microwave Irradiation and Denture Cleaning Agent. *Appl. Sci.* **2022**, *12*, 5860. https://doi.org/10.3390/app12125860

Academic Editor: Mary Anne Melo

Received: 8 May 2022
Accepted: 2 June 2022
Published: 9 June 2022

Abstract: The aim was to evaluate the disinfection of *E. coli*, *S. aureus* and *S. mutans* cultured on Polymethyl Methacrylate (PMMA) resin polymer using microwave disinfection and sodium perborate (DC). Biofilms of *E. coli*, *S. aureus* and *S. mutans* were cultured on the PMMA denture base for 24, 36, 48 and 96 h. Specimens were subjected to three disinfection protocols, including microwave disinfection in distilled water (MW-DW), sodium perborate (denture cleaning tablet) with distilled water (DC-DW) and a combination of MW-DC-DW for 1 to 5 min. Colony-forming units among the study groups were analyzed using Kruskal–Wallis and Mann–Whitney tests. For *E. coli* and *S. aureus* cultured on PMMA, the MW-DC-DW group displayed complete disinfection at 2 min of exposure. However, for both bacteria, the MW-DW disinfection group showed zero CFU at 3 min. DC disinfection for *E. coli* and *S. aureus* displayed zero CFU at 5 min of exposure ($p \leq 0.05$). For *S. mutans*, MW-DC-DW and MW-DW displayed zero CFU count at 1 min and 2 min, respectively. In DC-treated samples, CFU were significantly zero at 4 min when compared with the control at each growth time. A combination of MW irradiation with DC (sodium perborate) showed higher disinfection percentage of bacterial species on PMMA polymer denture bases compared to MW and DC alone. PMMA disinfection using DC displayed a lower antimicrobial disinfection percentage than the combined use of MW and DC as well as MW alone at 1 min or 2 min disinfection for *E. coli*, *S. aureus* and *S. mutans*.

Keywords: polymers; polymethyl methacrylate; bacteria; disinfection; oral health

1. Introduction

The artificial replacement of completely missing teeth with restoration of form, function and esthetics is critical to improving quality of life among edentulous patients [1]. A majority of edentulous patients experience psychosocial problems due to criticism, embarrassment and low confidence levels. For decades, different techniques and materials, i.e., wood, ivory and metals, have been employed for artificial denture fabrication [2]. Polymethyl methacrylate (PMMA) polymer denture resin material has been the material of choice for denture fabrication for the last 50 years, and has gained widespread acceptance due to its simplified processing and equipment, functional longevity, ease of maintenance, cost-effectiveness and esthetics [3].

The longer life expectancy of the elderly population leads to a higher number of individuals needing an artificial removable prosthesis [4]. Edentulous patients may lack manual dexterity and be unable to remove denture plaque. Therefore, PMMA denture patients are more susceptible to mucosal infections from bacterial and fungal species. In addition to various bacteria, *Candida albicans* (fungi) are most commonly associated with denture-related oral infections [5]. In order to overcome denture-related infections, different denture cleansers are used to effectively remove stains and food debris from the denture bases [6]. At present, different PMMA denture cleaning and disinfection techniques include mechanical, chemical and physical methods. Mechanical methods used for denture cleaning include the use of a toothbrush with water [7]. In their study, Veres et al. [8] revealed that 60–90% of denture patients practice mechanical cleaning in association with products such as toothpaste, soap or water. However, use of an unsuitable toothbrush with a dentifrice may also lead to surface roughness, which allows more microbial colonization [9]. In order to overcome drawbacks of mechanical cleaning, soaking PMMA denture bases in chemical disinfecting agents has gained more acceptance in clinical dentistry due to antimicrobial activity [10]. Chemical agents commonly used for denture cleaning include sodium hypochlorite, glutaraldehyde and chlorine dioxide, 0.12% Chlorhexidine gluconate, sodium perborate and alkaline peroxide [11]. The prosthesis is commonly immersed in the chemical agent for a specific duration for plaque removal and disinfection [11]. Effervescent tablets yielding an alkaline peroxide dilution with water are the preferred denture cleansers [12]. However, chemical disinfecting agents used for denture cleaning are known to damage acrylic resin, initiate surface staining or discoloration, and have been reported to be have cytotoxic ingredients [11].

Apart from mechanical and chemical means of denture disinfection, microwave oven disinfection is an effortless and inexpensive method for PMMA dentures [13,14]. Microwave irradiation allows for disinfection of not just bacteria and fungal species, but can also eliminate viruses [8]. Microwave irradiation can be a convenient, quick and cost effective household method for disinfecting PMMA dentures [13]. Studies have revealed that microwave disinfection has more potency for denture disinfection compared to sodium hypochlorite alone [8]. However, limited evidence is available to validate the disinfection protocol and standards for microwave irradiation to be a standard disinfection technique for acrylic dentures.

Previous studies investigating microwave disinfection of acrylic polymers have utilized high power (650 to 1400 watts) and increased exposure times (3 min, 5 min, 8, min, 10 min) [12–14]. However, these irradiation parameters can result in physical damage to polymeric denture base resin, including its dimensional stability and mechanical properties [15]. Therefore, it is critical to assess the low power and duration protocol for microwave irradiation in comparison to chemical disinfection for acrylic denture resin disinfection. It is hypothesized that the bacterial disinfection (cleaning efficacy) of PMMA denture base resin with a cleaning tablet or with microwave irradiation at low power (watt) and duration (1, 2 and 3 min) will be comparable. Therefore, the present study aimed to evaluate the disinfection efficacy of denture cleaning tablets (sodium perborate) and microwave irradiation at low power and duration (1, 2 and 3 min) for *E. coli*, *S. aureus* and *S. mutans* cultured on the PMMA denture base polymer.

2. Materials and Methods

In the present study, sample preparation was performed at the Dr Ishrat-ul-Ebad Institute of Oral Health Sciences (Department of Prosthodontics); however, the microbiological testing along with the disinfection interventions were undertaken at the Pakistan Council of Scientific & Industrial Research Laboratories. (PCSIR). The study assessed the antibacterial disinfection of PMMA denture resin polymer using individual and combined use of distilled water (DW), microwave disinfection (MW) and denture cleaning agent (DC). Disinfection of PMMA polymer was assessed on bacteria cultured at 24, 36, 48 and 96 h and with disinfecting durations of 1 to 5 min.

2.1. Specimen Preparation

Sample size was calculated using Pass version 11 (NCSS Statistical software, Kaysville, UT, USA), employing one-way ANOVA with 99% confidence interval, 99% power of the test, and with means and standard deviation of *C. Albicans* viable cells in groups: DW = 7.47×10^5, DC = 4.82×10^5, DW with MW 1 min = 4.49×10^5, DC with MW 1 min = 2.64×10^5, DC with MW 2 min = 0.00, DW with MW 3 min = 0.00, DC with M 3 min =0.00 [16]. Sample size was calculated with at least 10 specimens per sub-group (total 12 subgroups). However, a total of 168 specimens were fabricated ($n = 14$).

Test specimens ($n = 168$) of PMMA acrylic resin denture base were fabricated with the help of modelling wax (Yeti Dental GmbH, Engen, Germany) melted in a wax pot (Manfredi, San secondo di Pinerolo TO, Italy) and poured into a three-part preformed metal mold (30 × 30 × 15 mm). Wax patterns were impressed in a metallic denture flask filled with type III dental stone (Garrico Lab Stone, Heber Springs, AR, USA) to produce PMMA samples. De-waxing was performed using boiling water for 6 min. Heat polymerized PMMA acrylic resin was mixed and packed at a doughy stage according to manufacturer recommendations at a powder: liquid ratio, 2.3 g of polymer powder to 1 mL of liquid monomer (heat-cured acrylic provided by MR Dental, Plymouth, UK). A Hydraulic press was used for packing the denture base resin with a sheet of separating plastic between the two halves. Heat cure PMMA was polymerized in a thermostatically controlled water bath (Manfredi–Acrydig 12) and processed for 74 °C for two hours followed by 100 °C for one hour (Figure 1).

All specimens were allowed to cool at room temperature before de-flasking and were immersed in distilled water at room temperature for 48 h for residual monomer elimination. The excess resin was trimmed with a metal bur (Denfac Acrylic trimming burs). All PMMA specimens were wet-polished with abrasive paper (# 1200, water resistant) and buffed (Dialap ML150P; Maruto, Tokyo, Japan). Final finishing was performed with an alumina-based abrasive (particle size: 0.3 μm). Post-finishing, all specimens were autoclaved at 121 °C for 15 min. For the purpose of reliability, a single operator prepared all the specimens.

2.2. Study Groups

All specimens were divided into three main groups based on disinfection techniques, which are as follows.

Group MW-DW: microwave (MW) radiation was used to disinfect the specimens contaminated with a mixture of three isolates immersed in distilled water in a glass beaker. The glass container was placed in the microwave oven and the specimens were sterilized at 450 W. Based on duration of MW radiation, specimens were divided into MW-DW1 (1 min), MW-DW2 (2 min) and MW-DW3 (3 min). Different specimens in each subgroup were assessed at 24 h, 36 h, 48 h and 96 h, respectively.

Group DC-DW: specimens were immersed in distilled water with a denture cleaning tablet (sodium perborate) (DC) (Fittydent international, GmbH, Wien, Austria) added to it. Based on the duration of immersion in DC, the specimens were divided into DC-DW1 (1 min), DC-DW2 (2 min) and DC-DW3 (3 min). Different specimens in each subgroup were assessed at 24 h, 36 h, 48 h and 96 h, respectively.

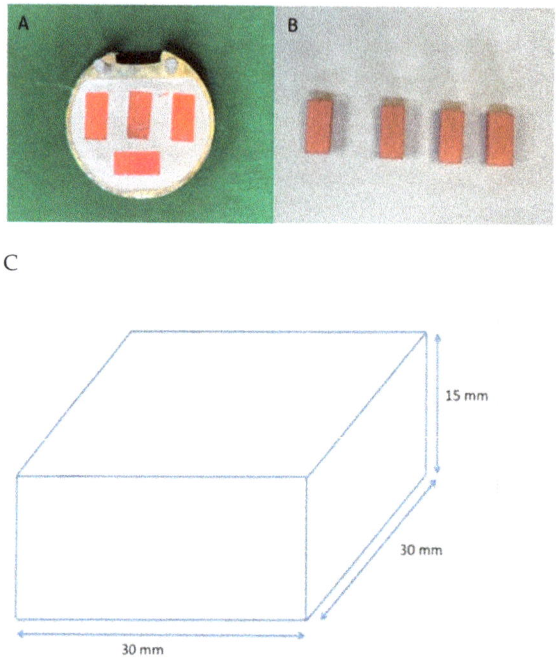

Figure 1. PMMA sample fabrication. (**A**) PMMA packing in plaster molds and (**B**) Finished PMMA samples. (**C**) Dimensions of each PMMA sample.

Group MW-DC-DW: the glass beaker containing 200 mL of distilled water in which a denture cleaning tablet was dissolved for five minutes was placed in the microwave oven and irradiated at 450 W. Based on duration of MW radiation, specimens were divided into MW-DC-DW1 (1 min), MW-DC-DW2 (2 min) and MW-DC-DW3 (3 min). The temperature of the solution was kept between 65 °C to 71 °C with ±2 °C. Different specimens in each subgroup were assessed at 24 h, 36 h, 48 h and 96 h, respectively. With five specimens in each subgroup, a total of 180 specimens in each disinfection group were employed.

Positive Control Group: in this group the acrylic resin specimens were immersed in glass beaker (250 mL size) containing 200 mL of distilled water at room temperature. The glass container was placed in the center of the microwave oven chamber (Samsung 2450 MHz, 800 W) but was not irradiated.

Negative Control Group: the purpose of this group was to establish the disinfection of the specimens and accuracy of the test. For each bacterium, sterilized specimens were placed in a container with sterilized water.

2.3. Biofilm Formation Assay

Overnight specimens of 0.1 mL of *E. coli* (ATCC# 8739), *S. aureus* (ATCC# 25923) and *S. mutans* (ATCC# 25175) subject isolates were inoculated into 100 mL of sterile Tryptic Soy Broth (TSB-Oxoid, Basingstoke, Hampshire, UK). The group contained a mixture of all three isolates. The sterile acrylic specimens were inoculated into each flask aseptically and incubated at 37 °C for 24 h, 36 h, 48 h and 96 h. After incubation, the acrylic specimens were collected and washed with distilled water to remove debris and loosely attached cells (Figure 2). Specimens were placed in phosphate buffered saline (PBS) (pH-7) and vortexed for two minutes. After vortexing, the specimens were, as has been explained, exposed to three treatment regimens in study groups (MW-DW, DC-DW and MW-DC-DW) at different exposure times. For selective isolation of bacteria, *E. coli*, *S. aureus* and *S. mutans* were cultured in Eosin Methylene Blue Agar (EMB Oxoid Basingstoke, Hampshire, UK) (Figure 3), Baird-Parker Agar with egg yolk tellurite (Fisher Scientific, Port Salvo, Portugal) (Figure 4) and Brain Heart Infusion (BHI) Broth (Difco, Detroit, MI, USA). The growth was monitored and CFU were counted.

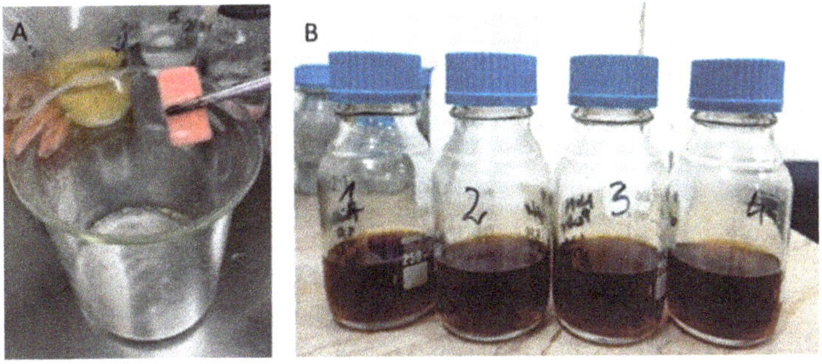

Figure 2. (**A**) Acquisition of PMMA samples after bacterial incubation. (**B**) Incubation of PMMA samples for 24 h, 36 h, 48 h and 96 h.

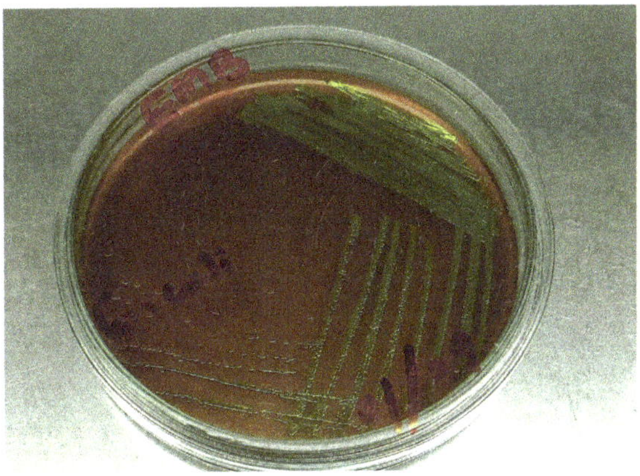

Figure 3. Pure culture of *E. coli* on EMB agar.

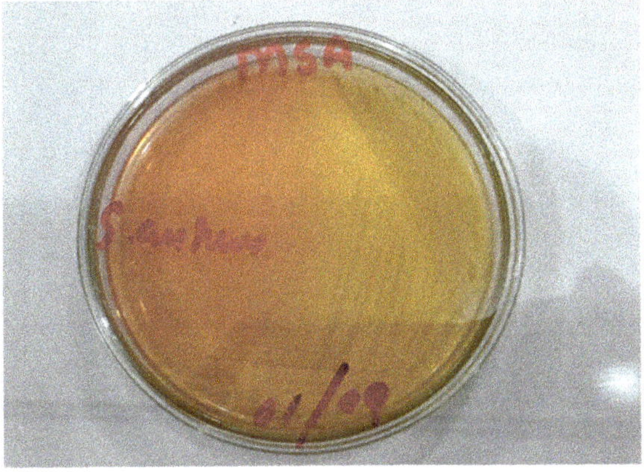

Figure 4. Pure culture of *S. aureus* on Mannitol Salt agar.

2.4. *Quantification of Biofilm*

The PMMA slides were collected after 24 h to 96 h and adhesion was assessed using crystal violet binding assay. Bacterial growth was fixed with acetic acid, followed by staining with 3% crystal violet (Ezzy Stain) and washing with PBS at 7.0 pH. Heat fixation was performed for 30 s followed by acetone washing. For accuracy of assessment and measurements, the procedure was repeated twice (per sample).

2.5. *Scanning Electron Microscopy*

Scanning electron microscopy was carried out for the analysis of the production of extracellular matrix material and to observe the bacterial presence and disinfection on acrylic slides. Biofilm slides were divided into 4 mm sections and washed with distilled water to remove the debris and were then negatively stained with 0.02% Uranyl acetate for 30 s. These 4 mm slides were platinum coated in a coating machine (JEOL 3000 FC, Tokyo, Japan), and the sections showed the presence of biofilm material when examined directly in SEM (JOEL-JSM IT-100, Tokyo, Japan).

2.6. *Statistical Analysis*

Data were entered and analyzed using SPSS (IBM Statistics software version 21, New York, NY, USA). For average comparison among exposure time, Kruskal–Wallis analysis was applied to each of the bacteria (*E. coli, S. aureus* and *S. mutan)* by treatment and growth time as CFU count were not normally distributed. The Mann–Whitney test was applied for pair-wise comparison of each exposure time with treatment and growth time controlled. A p-value of < 0.05 was considered as statistically significant.

3. Results

Mean CFU comparison for *E. coli* with the control for each disinfecting regime and growth time is presented in Table 1. From the results, it was found that all the specimens incubated at 24, 36, 48 and 96 h and exposed to MW-DW-DC and MW-DW disinfection for 1, 2 and 3 min demonstrated significant difference in mean CFU of *E. coli* compared to the control without disinfection. However, for the MW-DW-DC group, complete disinfection was achieved at 2 min of disinfection at any growth time (p-value < 0.05). Conversely, for MW-DW disinfection, CFU was significantly zero at 3 min when compared with control at any growth time (p-value $\leq$ 0.05). For DC-DW specimens, CFU levels were approximately zero at five minutes when compared with the control at different growth durations (p-value $\leq$ 0.05) (Table 1).

Mean CFU comparison for *S. aureus* among the control for each disinfecting regime and growth times is presented in Table 2. For MW-DW-DC, CFU were significantly zero at 2 min of disinfection when compared with control at any growth time (p-value $\leq$ 0.05). For MW-DW specimens, CFU were significantly zero at 3 min of disinfection when compared with control at each growth time (p-value $\leq$ 0.05). For DC-DW disinfection, CFU were significantly zero at five minutes when compared with control at each growth time (p-value $\leq$ 0.05) (Table 2).

Mean CFU comparison for *S. mutans* with the control for each disinfecting regime and growth time is presented in Table 3. For MW-DW-DC, CFU were significantly zero at 2 min when compared with control (p-value $\leq$ 0.05). For MW-DW disinfection, CFU were significantly zero at three mins when compared it with control at each growth time (p-value $\leq$ 0.05). For DC-DW, CFU were significantly zero at 5 min when compared with control at each growth time (p-value $\leq$ 0.05) (Table 3).

Table 1. Presenting Mean CFU difference among the test and control (non-disinfected) samples for *E. coli*.

Growth Time & Exposure Time vs. Control	Control CFU	MW-DC-DW	MW + DW	DC + DW
24 h	2254	MD (*p*-Value)	MD (*p*-Value)	MD (*p*-Value)
1 min		−2166 *	−1251*	−551
2 min		−2254 *	−2202 *	−1241 *
3 min		−2254 *	−2254 *	−1257 *
4 min		−2254 *	−2254 *	−2165 *
5 min		−2254 *	−2254 *	−2238 *
36 h	4409			
1 min		−4300 *	−2626 *	−404
2 min		−4409 *	−4363 *	−1395
3 min		−4409 *	−4409 *	−2310 *
4 min		−4409 *	−4409 *	−4323 *
5 min		−4409 *	−4409 *	−4398 *
48 h	1532			
1 min		−1359 *	−556 *	−208
2 min		−1532 *	−1459 *	−324
3 min		−1532 *	−1532 *	−530 *
4 min		−1532 *	−1532 *	−1465 *
5 min		−1532 *	−1532 *	−1524 *
96 h	179			
1 min		−156 *	−77 *	−94 *
2 min		−179 *	−118 *	−98 *
3 min		−179 *	−179 *	−107 *
4 min		−179 *	−179 *	−156 *
5 min		−179 *	−179 *	−176 *

* Significant at 5% using Mann–Whitney Test; MW: Microwave; DW: Distilled Water; MD: Mean Difference.

Table 2. Presenting Mean CFU difference among the test and control (non-disinfected) samples for *S. aureus*.

Growth Time & Exposure Time vs. Control	Control CFU	MW-DC-DW	MW + DW	DC + DW
24 h	1200	MD (*p*-Value)	MD (*p*-Value)	MD (*p*-Value)
1 min		−1119	−293 *	−194
2 min		−1200 *	−1165 *	−213
3 min		−1200 *	−1200 *	−399
4 min		1200 *	1200 *	−1144 *
5 min		1200 *	1200 *	−1200 *
36 h	3017			
1 min		−2819 *	−1928 *	−210
2 min		−3017 *	−2960 *	−1710 *
3 min		−3017 *	−3017 *	−2417 *
4 min		−3017 *	−3017 *	−2983 *
5 min		−3017 *	−3017 *	−3017 *
48 h	1175			
1 min		−994 *	−99	−87
2 min		−1175 *	−1124 *	−173
3 min		−1175 *	−1175 *	−188
4 min		−1175 *	−1175 *	−1095 *
5 min		−1175 *	−1175 *	−1175 *
96 h	187			
1 min		−166 *	−98 *	−98 *
2 min		−187 *	−144 *	−116 *
3 min		−187 *	−187 *	−118 *
4 min		−187 *	−187 *	−183 *
5 min		−187 *	−187 *	−187 *

* Significant at 5% using Mann–Whitney Test; MW: Microwave; DW: Distil Water; Tab: Tablet; MD: Mean Difference.

Table 3. Presenting Mean CFU difference among the test and control (non-disinfected) samples for *S. mutans*.

Growth Time & Exposure Time vs. Control	Control CFU	MW + DW + DC	MW + DW	DC + DW
24 h	115	MD (*p*-Value)	MD (*p*-Value)	MD (*p*-Value)
1 min		−115 *	−109 *	−20
2 min		−115 *	−115 *	−39
3 min		−115 *	−115 *	−71
4 min		−115 *	−115 *	−115 *
5 min		−115 *	−115 *	−115 *
36 h	199			
1 min		−199 *	−180 *	−103 *
2 min		−199 *	−199 *	−104 *
3 min		−199 *	−199 *	−177 *
4 min		−199 *	−199 *	−199 *
5 min		−199 *	−199 *	−199 *
48 h	81			
1 min		−81 *	−68 *	−14
2 min		−81 *	−81 *	−14
3 min		−81 *	−81 *	−38 *
4 min		−81 *	−81 *	−81 *
5 min		−81 *	−81 *	−81 *
96 h	46			
1 min		−46 *	−42 *	−6
2 min		−46 *	−46 *	−14
3 min		−46 *	−46 *	−25 *
4 min		−46 *	−46 *	−46 *
5 min		−46 *	−46 *	−46 *

* Significant at 5% using Mann–Whitney Test; MW: Microwave; DW: Distil Water; Tab: Tablet; MD: Mean Difference.

SEM analysis of the samples showed positive consortia-harboring cocci and bacilli (Figure 5). Using SEM, live cells of *E. coli* (rods), *S. aureus* and *S. mutans* (cocci) attached to the PMMA specimens were observed. After the disinfection procedure, dead cells were observed attached to the PMMA surface in the form of a small, agglomerated, mass of irregular-shaped cells (Figure 5).

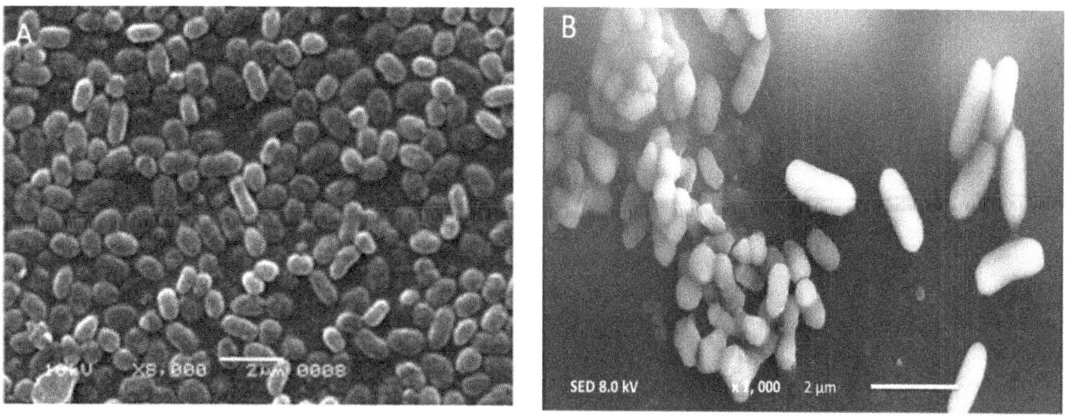

Figure 5. *Cont.*

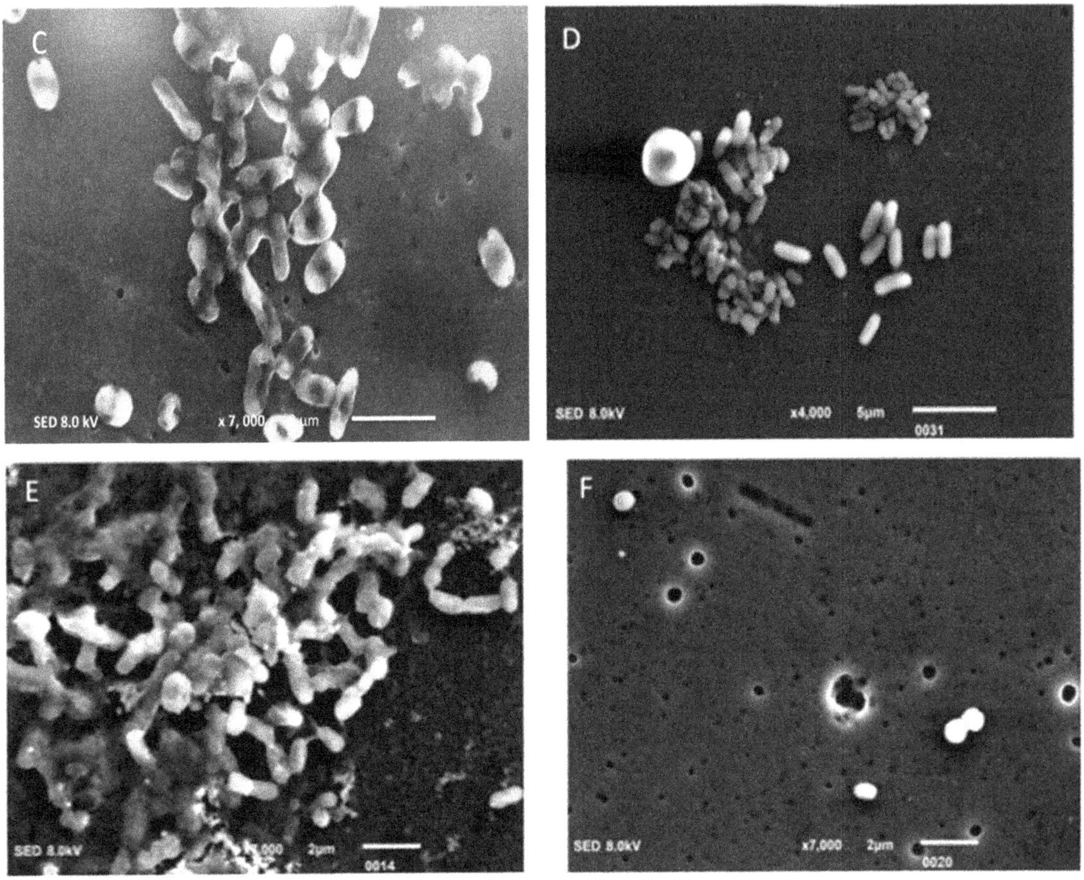

Figure 5. (**A**) Micrograph of biofilm positive consortia harboring cocci and bacilli. (**B**) Observe the attachment of mixture of live cells *E. coli* (rods) *S. aureus* and *S. mutans* (Cocci) on PMMA prior to disinfection. (**C**) Dead cells attached to the surface of acrylic slide, disinfected by exposure to MW at 450 W for 1 min (24 h). (**D**) Dead cells attached to the PMMA surface, disinfected by exposure to MW at 450 W for 1 min (48 h). (**E**) Dead cells on PMMA surface disinfected by MW with DC for 2 min (36 h). (**F**) PMMA surface disinfected with DW with DC for 5 min (24 h).

4. Discussion

The current study was conducted to evaluate the cumulative effect of microwave disinfection and a denture cleaning tablet (sodium perborate) with reduced time and power on the elimination of bacterial biofilm from PMMA denture base resin. It was based on the hypothesis that there will be no significant difference in cleaning efficacy of DC, MW and their combination on the disinfection of *E. coli*, *S. aureus* and *S. mutans* incubated on the PMMA denture base. Therefore, the postulated hypothesis was rejected, as all the tested groups displayed variable results of denture disinfection as compared to the control.

The possibility that a dental prosthesis can be a cause of infection and cross-contamination between patients and dental staff additionally stresses the need for disinfection [17]. Dental appliance disinfection poses a significant problem for clinicians, particularly with regards to removable dentures [18]. The results of the present study revealed that zero mean CFU was obtained after exposing PMMA to MW irradiation in combination with DC within 2 min (MW-DW-DC). In comparison, zero mean CFU for samples exposed to microwave alone (MW + DW) was obtained at 3 min. The one-minute exposure of acrylic specimen to MW irradiation showed 50% reduction in CFU for *E. coli*, 24% for *S. aureus* and 96% for *S. mutans*. Conversely, the combination of MW and DC resulted in 97% reduction of *E. coli* count, 96% for *S. aureus* and 100% for *S. mutans* after 1 min of exposure. Thus, a combination of MW irradiation with DC showed higher disinfection percentage of bacterial species in disinfection of PMMA polymer denture bases compared to MW and DC alone.

Considering the effectiveness of microwave disinfection, its role for denture decontamination has gained much attention in the past [19]. Multiple previous pieces of research have used variable time and power to ascertain its efficacy for denture disinfection [20]. Webb et al. first identified that an unmodified domestic microwave can perform denture disinfection [10]. In the present study, the efficacy of microwave irradiation on acrylic resin contaminated with bacterial biofilm at different growth hours (24–36–48–96 h) was investigated keeping the power 450 W and irradiation time 1, 2 and 3 min. From the results, it was found that exposure to microwave irradiation took 3 min to achieve complete CFU count zero, i.e., complete removal of *S. aureus*, *S. mutans* and *E. coli* from acrylic slides. Moreover, comparative analysis revealed that the increase in incubation period to 36 h resulted in an increased CFU count. However, this increase in count has not shown any impact on the disinfection capabilities of microwave radiation. These results are in line with the outcomes of the studies conducted by Baysan et al. and Dixon et al. [21,22]. These findings may be due to the fact that microwave irradiation alters structural integrity and cell membrane permeability [23]. It also has detrimental effects on cell metabolism, thus leading to bacterial cell death. In contrast to microwave disinfection of PMMA polymer bases, a combination of chemical disinfection (denture cleaning tablets) and microwaving effectively and efficaciously disinfects the material faster than microwaving alone, within 2 min of exposure time [24]. It is also important to note that the dentures in the present study were immersed in distilled water during microwave irradiation, a practice which has been a technique for eliminating microorganisms [25].

It was also observed that chemical disinfection using DC displayed lower antimicrobial disinfection percentage than the combined use of MW and DC and MW alone at 1 min or 2 min disinfection time for all three bacteria. In the present study, the denture cleaning tablet used was Fittydent (Fittydent international, GmbH, Wien, Austria) According to the manufacturer information, the tablets containing sodium perborate not only clean removable dentures but also have antibacterial ability against bad odor microbes, dissolve tough stains and help in plaque removal [26]. In case of bacterial biofilm, as the exposure time to the cleaning tablet increases the CFU count decreases, demonstrating the antimicrobial efficacy of the cleaning tablet used. There was 9% to 61% reduction in CFU count as the exposure increased from 1 to 3 min, and 96% to 100% of removal of CFU counts when it was exposed between 4 to 5 min. Thus, the cleaning tablet had more efficacy against bacterial biofilm when specimens were exposed for 4 to 5 min. The present result was in agreement with Silva et al., who evaluated the efficacy of commercial denture cleanser containing sodium perborate on the disinfection of acrylic specimens contaminated with *C. albicans* and *S. mutans* [11]. They reported that denture cleaning tablets have effective antimicrobial efficacy against streptococcus species as compared to *C. albicans* [27].

It is worth mentioning that the current study was conducted in vitro, with inherent limitations when simulating in vivo environment. A more clinically relevant study protocol could be an ex vivo model, including the disinfection of a patient's personal dentures in the oral cavity and experimental MW disinfection in the laboratory for CFU evaluation. It is pertinent to mention that denture-related stomatitis is commonly associated with *Candida albicans* on the denture surface and oral cavity [5]; therefore, further randomized controlled trials assessing the efficacy of microwave disinfection on the *Candida* species are warranted. In light of the findings within the study limitation, MW disinfection in combination with denture cleaning agents are recommended for PMMA denture disinfection in denture wearers.

5. Conclusions

Combination of MW irradiation with DC (sodium perborate) showed higher disinfection percentage of bacterial species on PMMA polymer denture bases compared to MW and DC alone. PMMA disinfection using DC displayed lower antimicrobial disinfection percentage than the combined use of MW and DC as well as MW alone at 1 min or 2 min disinfection for *E. coli*, *S. aureus* and *S. mutans*.

Author Contributions: Conceptualization, Z.C., S.M.R.K., S.M., M.A. and Z.A.M.; methodology, K.A.A., S.A., M.A.A., F.U.R.Q. and F.V.; software, K.A.A., F.V., M.A. and T.A.; validation, Z.C., S.M.R.K., S.M. and M.A.; formal analysis, S.A., Z.A.M., F.U.R.Q., M.A.A. and F.V.; resources, F.U.R.Q., T.A. and F.V; data curation, S.A., Z.A.M., Z.C., F.U.R.Q. and M.A.; writing—original draft preparation, F.V., T.A. and Z.C.; writing—review and editing, F.U.R.Q., S.A., K.A.A. and M.A. All authors have read and agreed to the published version of the manuscript.

Funding: This research was supported by Princess Nourah bint Abdulrahman University Researchers Supporting Project number (PNURSP2022R6), Princess Nourah bint Abdulrahman University, Riyadh, Saudi Arabia.

Institutional Review Board Statement: The study protocol was reviewed by the ethics and review committee of Dr. Ishrat-ul-Ebad Institute of Oral Health Sciences, Dow University of Health Sciences.

Informed Consent Statement: Informed consent was obtained and patients informed of the risks and benefits of the use of extracted teeth in the experiment.

Data Availability Statement: The data are available on contact from the corresponding author.

Conflicts of Interest: The author declares no conflict of interest.

References

1. Hayran, Y.; Sarikaya, I.; Aydin, A.; Tekin, Y.H. Determination of the effective anticandidal concentration of denture cleanser tablets on some denture base resins. *J. Appl. Oral Sci.* **2018**, *26*, e20170077. [CrossRef] [PubMed]
2. Marsh, P.D.; Percival, R.S.; Challacombe, S.J. The influence of denture-wearing and age on the oral microflora. *J. Dent. Res.* **1992**, *71*, 1374–1381. [CrossRef] [PubMed]
3. Berger, J.C.; Driscoll, C.F.; Romberg, E.; Luo, Q.; Thompson, G. Surface roughness of denture base acrylic resins after processing and after polishing. *J. Prosthodont. Implant. Esthet. Reconstr. Dent.* **2006**, *15*, 180–186. [CrossRef] [PubMed]
4. Busscher, H.J.; Weerkamp, A.H.; van der Mei, H.C.; Van Pelt, A.W.; de Jong, H.P.; Arends, J. Measurement of the surface free energy of bacterial cell surfaces and its relevance for adhesion. *Appl. Environ. Microbiol.* **1984**, *48*, 980–983. [CrossRef] [PubMed]
5. Alrabiah, M.; Alsahhaf, A.; Alofi, R.S.; Al-Aali, K.A.; Abduljabbar, T.; Vohra, F. Efficacy of photodynamic therapy versus local nystatin in the treatment of denture stomatitis: A randomized clinical study. *Photodiagnosis Photodyn. Ther.* **2019**, *28*, 98–101. [CrossRef] [PubMed]
6. de Freitas Fernandes, F.S.; Pereira-Cenci, T.; Da Silva, W.J.; Filho, A.P.R.; Straioto, F.G.; Cury, A.A. Efficacy of denture cleansers on Candida spp. biofilm formed on polyamide and polymethyl methacrylate resins. *J. Prosthet. Dent.* **2011**, *105*, 51–58. [CrossRef]
7. Baba, Y.; Sato, Y.; Owada, G.; Minakuchi, S. Effectiveness of a combination denture-cleaning method versus a mechanical method: Comparison of denture cleanliness, patient satisfaction, and oral health-related quality of life. *J. Prosthodont. Res.* **2018**, *62*, 353–358. [CrossRef]
8. Veres, E.; Wolfaardt, J.; Hnizdo, E. Denture cleansers: Part III—A survey of materials and methods employed by denture wearers. *J. Dent. Assoc. S. Afr.* **1985**, *40*, 591–594.
9. Nishi, Y.; Seto, K.; Kamashita, Y.; Take, C.; Kurono, A.; Nagaoka, E. Examination of denture-cleaning methods based on the quantity of microorganisms adhering to a denture. *Gerodontology* **2012**, *29*, e259–e266. [CrossRef]
10. Webb, B.; Thomas, C.; Harty, D.; Willcox, M. Effectiveness of two methods of denture disinfection. *J. Oral Rehabil.* **1998**, *25*, 416–423. [CrossRef]
11. Silva, M.M.; Vergani, C.E.; Giampaolo, E.T.; Neppelenbroek, K.H.; Spolidorio, D.M.P.; Machado, A.L. Effectiveness of microwave irradiation on the disinfection of complete dentures. *Int. J. Prosthodont.* **2006**, *19*, 288–293. [PubMed]
12. Rohrer, M.D.; Bulard, R.A. Microwave disinfection. *J. Am. Dent. Assoc.* **1985**, *110*, 194–198. [CrossRef] [PubMed]
13. Pavan, S.; Arioli Filho, J.N.; Santos, P.H.d.; Mollo, F.d.A., Jr. Effect of microwave treatments on dimensional accuracy of maxillary acrylic resin denture base. *Braz. Dent. J.* **2005**, *16*, 119–123. [CrossRef] [PubMed]
14. Sanita, P.V.; Vergani, C.E.; Giampaolo, E.T.; Pavarina, A.C.; Machado, A.L. Growth of Candida species on complete dentures: Effect of microwave disinfection. *Mycoses* **2009**, *52*, 154–160. [CrossRef] [PubMed]
15. Klironomos, T.; Katsimpali, A.; Polyzois, G. The effect of microwave disinfection on denture base polymers, liners and teeth: A Basic overview. *Acta Stomatol. Croat.* **2015**, *49*, 242–253. [CrossRef] [PubMed]

16. Senna, P.M.; Sotto-Maior, B.S.; Silva, W.J.; Del Bel Cury, A.A. Adding denture cleanser to microwave disinfection regimen to reduce the irradiation time and the exposure of dentures to high temperatures. *Gerodontology* **2013**, *30*, 26–31. [CrossRef]
17. Arita, M.; Nagayoshi, M.; Fukuizumi, T.; Okinaga, T.; Masumi, S.; Morikawa, M.; Kakinoki, Y.; Nishihara, T. Microbicidal efficacy of ozonated water against Candida albicans adhering to acrylic denture plates. *Oral Microbiol. Immunol.* **2005**, *20*, 206–210. [CrossRef]
18. Meşe, A.; Meşe, S. Effect of microwave energy on fungal growth of resilient denture liner material. *Biotechnol. Biotechnol. Equip.* **2007**, *21*, 91–93. [CrossRef]
19. Tarbet, W.J. Denture plaque: Quiet destroyer. *J. Prosthet. Dent.* **1982**, *48*, 647–652. [CrossRef]
20. Dills, S.; Olshan, A.; Goldner, S.; Brogdon, C. Comparison of the antimicrobial capability of an abrasive paste and chemical-soak denture cleaners. *J. Prosthet. Dent.* **1988**, *60*, 467–470. [CrossRef]
21. Baysan, A.; Whiley, R.; Wright, P.S. Use of microwave energy to disinfect a long-term soft lining material contaminated with Candida albicans or Staphylococcus aureus. *J. Prosthet. Dent.* **1998**, *79*, 454–458. [CrossRef]
22. Dixon, D.L.; Breeding, L.C.; Faler, T.A. Microwave disinfection of denture base materials colonized with Candida albicans. *J. Prosthet. Dent.* **1999**, *81*, 207–214. [CrossRef]
23. Kulak, Y.; Arikan, A.; Kazazoglu, E. Existence of Candida albicans and microorganisms in denture stomatitis patients. *J. Oral Rehabil.* **1997**, *24*, 788–790. [CrossRef] [PubMed]
24. Campanha, N.H.; Pavarina, A.C.; Brunetti, I.L.; Vergani, C.E.; Machado, A.L.; Spolidorio, D.M. Candida albicans inactivation and cell membrane integrity damage by microwave irradiation. *Mycoses* **2007**, *50*, 140–147. [CrossRef]
25. Culkin, K.A.; Fung, D.Y. Destruction of Escherichia coli and Salmonella typhimurium in microwave-cooked soups. *J. Milk Food Technol.* **1975**, *38*, 8–15. [CrossRef]
26. McDonnell, G.; Russell, A.D. Antiseptics and disinfectants: Activity, action, and resistance. *Clin. Microbiol. Rev.* **1999**, *12*, 147–179. [CrossRef]
27. Block, S.S. *Disinfection, Disinfection, and Preservation*; Lippincott Williams & Wilkins: Philadelphia, PA, USA, 2001.

applied sciences

MDPI

Article

Oral Health in Migrants: An Observational Study on the Oral Health Status of a Migrant Cohort Coming from Middle- and Low-Income Countries

Dorina Lauritano [1,*,†], Giulia Moreo [2,†], Marcella Martinelli [3], Vincenzo Campanella [4], Claudio Arcuri [4] and Francesco Carinci [1]

1 Department of Translational Medicine and for Romagna, University of Ferrara, 44121 Ferrara, Italy; crc@unife.it
2 Dental and Maxillo-Facial Surgery Unit, IRCCS Ca' Granda Ospedale Maggiore Policlinico di Milano, 20122 Milan, Italy; giulia.moreo@unimi.it
3 Department of Experimental, Diagnostic and Specialty Medicine, University of Bologna, 40126 Bologna, Italy; marcella.martinelli@unibo.it
4 Department of Clinical Science and Translational Medicine, University of Rome "Tor Vergata", 00113 Rome, Italy; vincenzo.campanella@uniroma2.it (V.C.); claudio.arcuri@uniroma2.it (C.A.)
* Correspondence: dorina.lauritano@unife.it
† These authors contributed equally to this work.

Abstract: *Introduction.* The oral health conditions of migrants coming from middle- and low-income countries to developed countries have been shown to be poorer than those of the host population. Since the phenomenon of migration has continued to grow over the past five decades, the oral health status of migrant populations should be an issue of great concern. *Objectives.* The objective of our observational study was to analyse the periodontal health status and the prevalence of dental caries and lesions of the oral mucosa in a migrant cohort resident in Italy, assessing the association between the oral health status and the educational level of the included patients. *Materials and Methods.* Our research was conducted at the dental department of Policlinico Tor Vergata (Rome, Italy). A sample of 200 migrants coming from middle- and low-income countries, aged between 3 and 37, was included in our study. Each patient underwent a physical examination of the oral cavity, recording the DMFT/dmft index, Community Periodontal Index of Treatment Needs (CPI), and lesions of the oral mucosa. The one-way ANOVA test was used to establish the correlation between the oral health status and the educational level of the participants. *Results.* Many participants (62.5%) showed a DMFT/dmft Index ≥ 4; only 27% of the migrants had a DMFT/dmft Index lower than 4, and only 21 of them (10.5%) were recorded at 0. A CPI equal to 0 or 1 was observed in 131 patients (65.5%), while only 30 participants presented a CPI equal to or higher than 4 (15%), and 19.5% (39 patients) were assigned to code 2 and 3. Significant statistical differences were found in the CPI after adjusting data for the educational level of the included participants (p-value < 0.01). *Conclusions.* The data obtained in our research highlighted poor oral conditions among the analysed migrant population, recording a high prevalence of dental caries and inadequate oral hygiene habits.

Keywords: migrants; middle-and low-income countries; oral health of migrants; oral hygiene habits; migrant children's oral health status; dental caries; gingival bleeding

Citation: Lauritano, D.; Moreo, G.; Martinelli, M.; Campanella, V.; Arcuri, C.; Carinci, F. Oral Health in Migrants: An Observational Study on the Oral Health Status of a Migrant Cohort Coming from Middle- and Low-Income Countries. *Appl. Sci.* **2022**, *12*, 5774. https://doi.org/10.3390/app12125774

Academic Editor: Bruno Chrcanovic

Received: 14 May 2022
Accepted: 5 June 2022
Published: 7 June 2022

1. Introduction

The literature has reported that a migration background can affect the oral health status of individuals [1,2], and, since the number of international migrants has continued to grow over the past five decades, this issue is salient.

According to the World Migration Report 2022, it is estimated that 218 million people now in a country other than their country of birth, and it was observed that the number of migrants is three times higher than in 1970 [3]. The International Migration Report by the United Nations [4] highlighted that in 2017 more than 50% of the total number of international migrants settled in developed countries.

It has been reported that 1.9 million persons migrated to the European Union (EU) from non-EU countries in 2020, and, in 2021, 5.3% of the entire population living in the EU were migrants [5].

It has been largely demonstrated that poor oral health can have a negative effect on general health, consistently impacting daily activities [6–10], so much so that oral health was defined as a Leading Health Indicator 2020 [11].

Many developing countries, such as Bangladesh, Sri Lanka, Pakistan, and India, presented a very low oral health literacy [12], which the World Health Organization defined as "the cognitive and social skills which determine the motivation and ability of individuals to gain access to, understand and use information in ways which promote and maintain good health" [13]. Furthermore, the healthcare system in these countries is often less qualified, not well organized, and without specific regulations [1].

Migrants move from their countries for many reasons, such as conflicts, persecution, and poverty [14,15], and they face many barriers once they reach the host country: social and economic inequalities, language and legal status problems, different cultural habits, and difficulties in accessing healthcare facilities [16,17]. Therefore, the phenomenon of migration, which includes economic, social, and emotional disruption, may negatively affect the quality of life of migrants and, consequently, their general health and oral health [18].

Several authors have demonstrated that people who migrated from middle- and low-income countries to high-income countries presented a lower oral health status, a condition due to cultural habits, religiosity, and social behaviours of their native nations [19–21].

Pabbla et al. [22] recorded in his systematic review a significantly higher prevalence of dental caries and gingival bleeding among migrant children coming from Asian and African countries compared to the population of the host countries (European countries). The same study highlighted a general poor knowledge of oral health among migrants and a lack of monitoring by migrant parents of the oral health of their children.

Serna et al. [23] investigated the dental care utilization among Hispanic migrant farmworkers in South Florida, showing that most of them did not have a past year dental visit, and they presented poor oral conditions, having difficulties in accessing healthcare services.

The article by Svensson et al. [24] assessed poor oral hygiene in half of the analysed migrant children coming from Somalia to Sweden, underlining that 78–82% of them had never visited a dental department in their native nation, and that, in most cases, parents did not assist their youngest children during tooth cleaning.

There are different risk factors, which prevent migrant children from maintaining good oral health, and Reza et al. [25] grouped them into 3 levels: (1) the child level, due to the different oral hygiene habits, (2) the family level, including economic barriers, parents' inadequate knowledge of oral health practices, and scant attention to the oral health condition of their children and (3) the community level, that is, lack of medical insurance.

Considering these, interventions aimed at reducing disparities in access to dental treatment are necessary [26–28].

This observational study aimed to assess the periodontal health status and the prevalence of dental caries and lesions of the oral mucosa of a migrant cohort coming from middle- and low-income countries to Italy. Furthermore, we investigated the association between dental caries, periodontal status, and educational level of the included migrants.

2. Materials and Methods

A cross-sectional design was used to realize this observational study, recording the prevalence of dental caries, oral mucosa lesions, and gingival health status in a migrant population coming from middle- and low-income countries and resident in Italy. Our research included 200 migrant subjects, aged between 3 and 37, who visited the dental department of the Policlinico Tor Vergata (Rome, Italy) in the period from 1 September

2021 to 31 March 2022. This study was approved by the Fondazione Policlinico Tor Vergata (Rome, Italy) ethical committee (approval number N. 0001919/2020 del 29/01/2020)

Inclusion criteria were as follows: (a) subjects under the age of 40 years; (b) patients identified as migrants; (c) persons whose country of origin was identified as middle- or low-income country; (d) participants aged 18 years and older had to be born in middle- or low-income countries; (e) parents of patients under the age of 18 had to be born in middle- or low-income countries; (f) migrants resident in Italy. Persons aged 40 years and older, migrants identified as asylum seekers or refugees, non-migrant subjects, or migrants coming from high-income countries were excluded from our study.

The selected sample included 110 female and 90 male subjects aged between 3 and 37 (median age 9.39 years old ± 6.49). This research selected 188 children, of whom 30 were born in middle- and low-income countries, and 158 were born in Italy but had both parents whose country of birth was a middle- or low-income country. The subjects selected for our study came from (Table 1):

Table 1. Country of origin of included patients.

Country	Total N.	Parents' Country of Birth N.	Child's Country of Birth (Aged between 3 and 17) N.	Adult Subject's Country of Birth (Aged between 18 and 37) N.
Egypt	42	28	11	3
Peru	36	28	7	1
Ecuador	13	12	/	1
The Philippines	12	11	/	1
Tunisia	4	2	1	1
Morocco	10	9	/	1
Montenegro	3	3	/	/
Bangladesh	5	4	1	/
Pakistan	1	/	1	/
Algeria	1	1	/	/
Romania	13	10	1	2
Sri Lanka	6	5	1	/
Nepal	1	1	/	/
Moldavia	5	4	1	/
Albania	13	11	/	2
Turkey	2	2	/	/
Kosovo	2	2	/	/
Senegal	3	3	/	/
Bolivia	1	1	/	/
Venezuela	1	/	/	1
Bosnia and Herzegovina	1	/	1	/
India	3	3	/	/
El Salvador	7	4	3	/
China	6	6	/	/
Ukraine	6	4	2	/
Cameroon	1	1	/	/
Mauritius	1	1	/	/
Saudi Arabia	1	1	/	/
Total	200	158	30	12

Africa (n = 62): Egypt (n = 42), Morocco (n = 10), Tunisia (n = 4), Senegal (n = 3), Algeria (n = 1), Cameroon (n = 1), and Mauritius (n = 1);

Central and South America (n = 58): Peru (n = 36), Ecuador (n = 13), El Salvador (n = 7), Venezuela (n = 1), and Bolivia (n = 1);

Europe (n = 45): Romania (n = 13), Albania (n = 13), Ukraine (n = 6), Moldavia (n = 5), Kosovo (n = 2), Montenegro (n = 3), Bosnia and Herzegovina (n = 1), and Turkey (n = 2);

Asia (n = 35): China (n = 6), India (n = 3), the Philippines (n = 12), Pakistan (n = 1), Sri Lanka (n = 6), Bangladesh (n = 5), Nepal (n = 1), and Saudi Arabia (n = 1).

2.1. Clinical Methods

Each patient underwent a physical examination of the oral cavity, during which the following clinical parameters were analysed:

Decayed Missing Filled Teeth Index for permanent dental elements/decayed missing filled teeth index for primary dental elements (DMFT/dmft) [29–31]. The presence of caries was determined following the International Caries Classification and Management System [32].

Community Periodontal Index of Treatment Needs (CPI) [33,34]:

Code 0 → no treatment needs;

Code 1 → bleeding gingiva after gentle probing;

Code 2 → periodontal pockets not deeper than 3 mm, presence of supra- or subgingival calculus or plaque;

Code 3 → 4–5 mm deep periodontal pockets;

Code 4 → 6 mm or deeper periodontal pockets.

Lesions of oral mucosa or new mucosal formations [35] with the following characteristics:

Acute/chronic onset ulcerative vesciculo-bullous lesions;

Red, blue, or red-purple plan/exophytic lesions, not removable white lesions, white and red, yellowish, pigmented lesions;

Swelling on palate, tongue, oral floor, vestibular mucosa, and/or gingiva.

The examination was conducted by one clinician, who performed the inspection of dental elements and the palpation of soft tissues (lips, tongue, cheeks, palate, and oral floor) of each participant. Periodontal probing was performed by the same operator using a probe with calibrations given in 3 mm sections. In order to diagnose the presence of caries, the operator used a dental mirror and a probe, analysing the occlusal, distal, mesial, palatal, and buccal portion of each dental element.

Aside from the clinical parameters, demographic data were collected: age, sex, country of birth, and level of education. In the case of minor subjects, the parents' level of education was taken into consideration. All these demographic data were recorded by using a questionnaire.

The medical history of the included patients was collected before proceeding with the oral examination, and each person's data were collected after obtaining informed consent.

2.2. Statistical Methods

The prevalence of dental caries (DMFT/dmft Index), oral mucosa lesions, and gingival health status (CPI) was obtained using means ± standard deviation and percentages. The calculation was performed according to country of origin, age group, and sex. The one-way ANOVA test was performed to analyse the correlation between the educational level and the clinical parameters (DMFT/dmft Index and CPI).

3. Results

Our study analysed the prevalence of dental caries, gingival health status, and lesions of oral mucosa in a sample of 200 migrants from middle- and low-income countries. Most participants (62.5%, 124 migrants out of 200) showed a DMFT/dmft Index $\geq$ 4; 27% of the migrants had a DMFT/dmft Index lower than 4, and only 21 of them (10.5%) were recorded at 0.

A CPI equal to 0 or 1 was observed in 131 patients (65.5%), while only 30 participants presented a CPI equal or higher than 4 (15%), and 19.5% (39 patients) of them were assigned code 2 and 3. The CPI results were low in the selected population.

The DMFT/dmft Index and CPI were on average 5.2 ± 2.23 and 1.4 ± 1.24, respectively (Table 1). Migrants born in Turkey, Pakistan, and Saudi Arabia had the highest mean values of the DMFT/dmft Index: 9 ± 3, 8, and 8, respectively. Romania had the highest CPI mean value (1.76 ± 2.32), while patients born in Senegal, Mauritius, Bolivia, Nepal, and Bosnia and Herzegovina had no bleeding after gentle probing (CPI = 0).

Higher mean values of DMFT/dmft Index were recorded in migrants coming from Bangladesh (7.6 ± 3.38), China (7.16 ± 3.43), Egypt (6.11 ± 4.06), Romania (6 ± 3.96), the Philippines (5.41 ± 4.03), El Salvador (5.42 ± 4.74), Albania (5.15 ± 3.5), and Ukraine (5.16 ± 2.79) than in those coming from the remaining countries.

The lowest DMFT/dmft Index mean value was found in migrants whose country of origin was Sri Lanka (1.5 ± 1.5).

Migrants coming from Africa and from Europe were found to have the lowest DMFT/dmft index (3.8 ± 1.13) and CPI (1.27 ± 0.62), respectively.

Asian countries (China, the Philippines, India, Pakistan, Sri Lanka, Bangladesh, Nepal, and Saudi Arabia) were recorded to have the highest DMFT/dmft index (6.83 ± 2.82) and a CPI equal to 1.58 ± 1.26. Participants born in Europe (Romania, Albania, Kosovo, Moldavia, Montenegro, Bosnia and Herzegovina, Ukraine, and Turkey) had a DMFT/dmft index higher than those born in Central and South America (Peru, Ecuador, Venezuela, Bolivia, and El Salvador): 5.36 ± 1.82 and 4.76 ± 0.92, respectively (Figures 1–4).

Patients whose country of origin was Africa (Egypt, Morocco, Tunisia, Algeria, Senegal, Cameroon, and Mauritius) showed a lower CPI (1.32 ± 1.3) than those coming from Central and South America (1.75 ± 1.71).

Based on the age range (Table 2), children aged between 3 and 6 years (n = 74) had the highest dmft index (6.5 ± 4.1). Adult subjects aged between 21 and 37 years (n = 14) had the highest CPI (3.4 ± 2.77). The DMFT/dmft Index in males was higher than in females (5.61 ± 4.10 and 4.5 ± 3.12, respectively), but the CPI was lower in males than in females (1.15 ± 1.67 and 1.45 ± 2.15, respectively) (Table 2). The mean value of the CPI in children and adolescents (3–6/7–12 and 13–20 years old) was equal to 1.01 ± 0.33, showing better periodontal health compared to the enrolled adult patients.

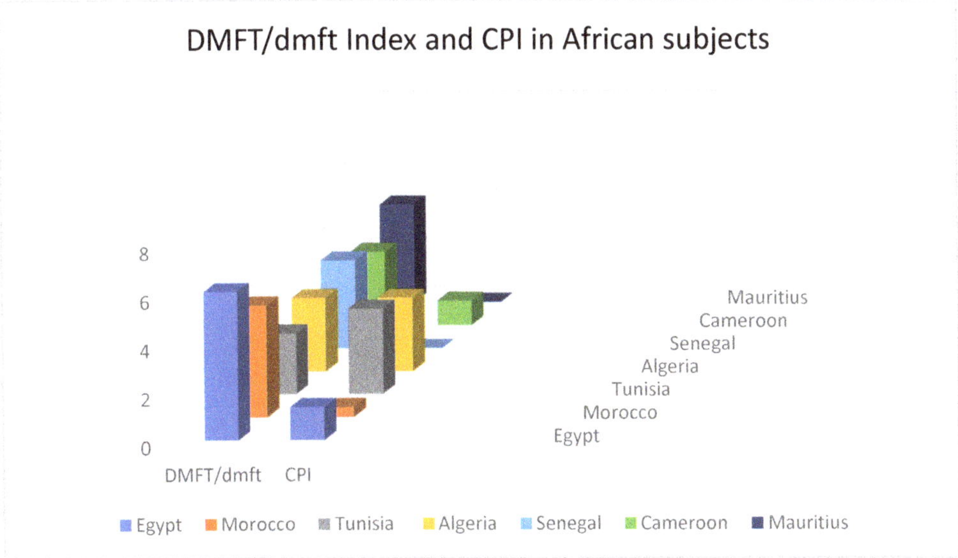

Figure 1. DMFT/dmft Index and CPI (means) in African patients.

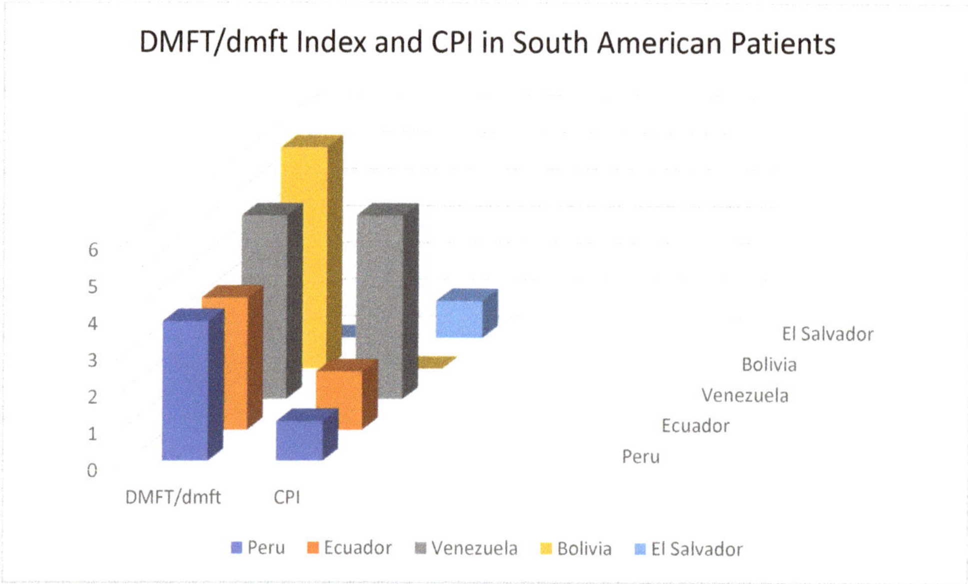

Figure 2. DMFT/dmft Index and CPI (means) in Central and South American patients.

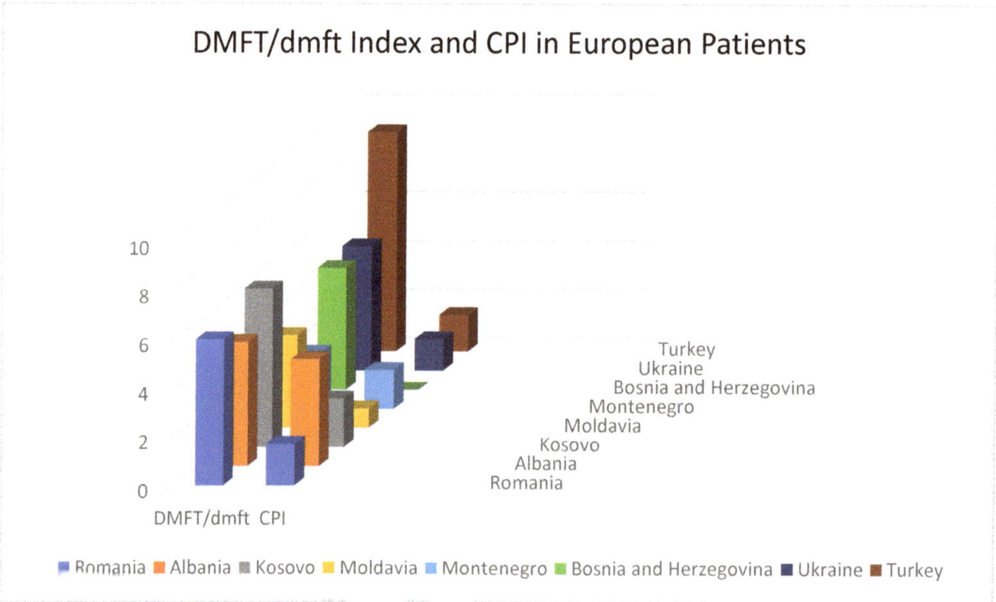

Figure 3. DMFT/dmft Index and CPI (means) in European patients.

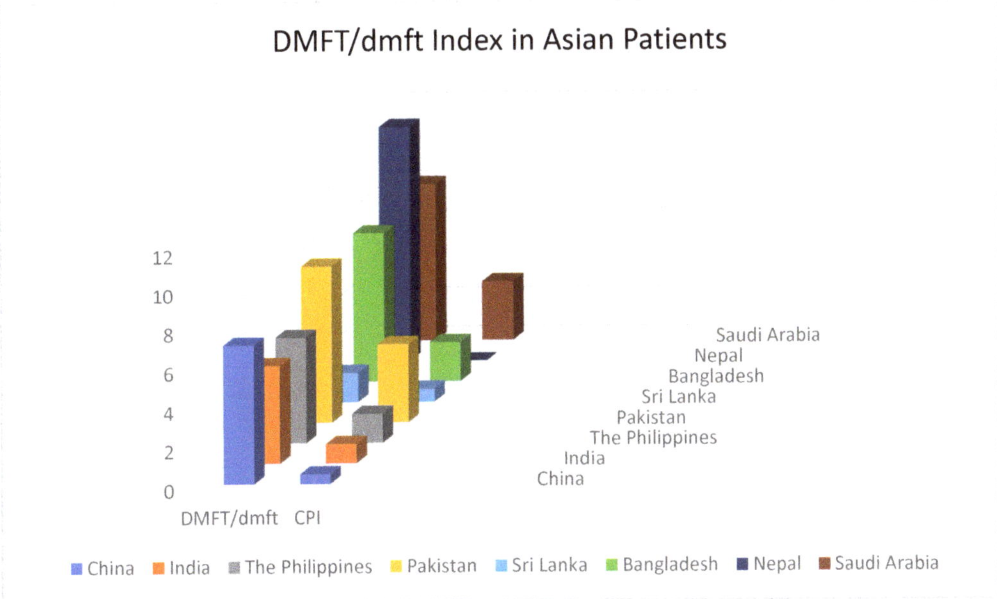

Figure 4. DMFT/dmft Index and CPI (means) in Asian patients.

Table 2. DMFT/dmft Index and CPI: Mean (M) and Standard Deviation (SD) according to the Country of Origin.

Country of Birth	Population Sample	DMFT/dmft (M)	DMFT/dmft (SD)	CPI (M)	CPI (SD)
Egypt	42	6.11	4.06	1.35	2.12
Morocco	10	4.6	4.29	0.4	0.66
Tunisia	4	2.5	1.5	3.5	2.29
Algeria	1	3	-	3	-
Senegal	3	3.66	0.94	0	0
Cameroon	1	3	-	1	-
Mauritius	1	4	-	0	-
Peru	36	3.805	2.83	1.08	1.73
Ecuador	13	3.61	2.09	1.69	2.39
Venezuela	1	5	-	5	-
Bolivia	1	6	-	0	-
El Salvador	7	5.42	4.74	1	1.60
The Philippines	12	5.41	4.03	1.5	1.89
China	6	7.16	3.43	0.5	1.11
India	3	5	1.41	1	1.41
Sri Lanka	6	1.5	1.5	0.66	1.49
Nepal	1	12	-	0	-
Albania	13	5.15	3.50	1.53	1.94
Romania	13	6	3.96	1.76	2.32
Ukraine	6	5.16	2.79	1.33	1.97
Moldavia	5	3.8	1.93	0.8	1.6
Turkey	2	9	3	1.5	1.5
Montenegro	3	2.33	0.47	1.66	2.35
Bosnia and Herzegovina	1	5	-	0	-
Kosovo	2	6.5	2.5	2	2
Pakistan	1	8	-	4	-
Bangladesh	5	7.6	3.38	2	1.26
Saudi Arabia	1	8	-	3	-
Total	200	5.28	2.23	1.4	1.24

Only 32 migrants out of 200 had attended university (DMFT/dmft mean value = 4.7 ± 2.8, CPI = 0.8 ± 1.5), 35 subjects had a secondary school degree (DMFT/dmft mean value = 5.7 ± 3.4,

CPI mean value = 2.1 ± 2.4), while most of the included migrants had achieved a high school degree (DMFT/dmft mean value = 4.8 ± 3.7, CPI mean value = 1.1 ± 1.8) (Table 3).

Table 3. DMFT/dmft Index and CPI: Mean (M) and Standard Deviation (SD) according to Age, Sex, and Education Level.

Age Range	Population Sample	DMFT/dmft (M)	DMFT/dmft (SD)	CPI (M)	CPI (SD)
3–6 years old	74	6.5	4.1	0.94	1.57
7–12 years old	98	4.6	3.01	1.45	2.20
13–20 years old	14	1.46	1.54	0.64	1.58
21–37 years old	14	3.8	2.45	3.4	2.77
Female	110	4.5	3.12	1.45	2.15
Male	90	5.61	4.10	1.15	1.67
Secondary School	35	5.7	3.4	2.1	2.4
High School	132	4.8	3.7	1.1	1.8
University	32	4.7	2.8	0.8	1.5

Significant statistical differences were found in the CPI after adjusting data for the educational level of the included participants (p-value < 0.01); on the contrary, the results regarding the DMFT/dmft Index were not significant at p-value < 0.01 (Table 4).

Table 4. DMFT/dmft Index and CPI adjusted for Educational Level: ANOVA Test.

Educational Level	DMFT/dmft (M)	DMFT/dmft (SD)	SS	df	MS	F	p-Value
Secondary School	5.7	3.4					
High School	4.8	3.7	23.57	2	11.78	0.92632 *	0.397728 *
University	4.7	2.8					
	CPI (M)	CPI (SD)	SS	df	MS	F	p-Value
Secondary School	2.1	2.4					
High School	1.1	1.8	35.90	2	17.95	4.85511 **	0.00874 **
University	0.8	1.5					

* Results NOT significant at p value $p < 0.01$. ** Results significant at p value $p < 0.01$.

None of the participants presented lesions of the oral mucosa.

4. Discussion

This study assessed the prevalence of dental caries and the health status of periodontal tissues among 200 migrants coming from developing countries and living in Northern Italy. To investigate the presence of dental caries we calculated the Decayed Missing Filled Teeth Index and decayed missing filled index for permanent and primary teeth, respectively. The DMFT/dmft represents the most important index used in epidemiological studies regarding the health status of the population, defining the number of decayed teeth, treated teeth, and the number of teeth missing due to carious lesions. It allows evaluation of the need for oral health interventions and strategies, to prevent dental decay [29,30]. The Community Periodontal Index of Treatment Needs was proposed in 1977 by the World Health Organization to measure periodontal disease and evaluate the treatment needs of populations [35–37].

Data obtained from the analysis reported, in general, a high prevalence of dental caries among the included patients, since more than half of them presented a DMFT/dmft ≥ 4, and only 10.5% had no decayed, filled, or missing teeth. Most of the selected sample was represented by children (age range 3–17 years old). Most of them were born in Italy but had both parents whose nation of birth is considered a middle- or low-income country.

In order to compare our data with a control population, we took into consideration the epidemiological study by Severino et al. [38]: this research evaluated the presence of Early Childhood Caries (ECC), by analysing the dmft index, in a paediatric Italian population composed of 76 children and aged up to 6 years. In the same study, the educational level of the parents was collected by using a questionnaire, recording that more than 36% and 22% of mothers and fathers, respectively, had attended college (university). The authors of this article reported that more than half of the selected patients (59.21%) had never experienced caries, and that the majority of the parents used to brush their teeth 3 or more than 3 times a day.

The control group for children older than 6 years could be represented by the population sample selected in the study by Campus et al. [39], in which the DMFT/dmft Index and the CPI were calculated in 5342 Italian children aged 12 years. The national mean DMFT was equal to 1.09% (95% CI 0.98–1.21), and 23.8% of the patients had gingival bleeding, while 28.7% had calculus.

On the basis of the data collected from the literature [38,39], and according to the results of our observational study, it can be stated that the prevalence of dental caries appears to be higher in the migrant subjects living in Italy and coming from middle- and low-income countries than in the native population.

Our results are in line with the data recorded by Ferrazzano et al. [40]: the authors of this paper studied the DMFT index among migrant children with low incomes in South Italy, demonstrating higher levels of dental caries compared to the non-migrant population and showing a higher Unmet Restorative Treatment Needs index in migrant children than that of the native population.

van Meijeren-van Lunteren et al. [41] analysed the oral health-related quality of life (OHRQoL) in migrant children (coming from Morocco, Indonesia, Suriname, and Turkey) resident in the Netherlands. This research highlighted that the prevalence of caries-free dentition was higher among native children than among migrant participants and suggested that the low OHRQoL was mediated by the oral health status and the socioeconomic position.

According to the data obtained in our research, higher mean values of DMFT/dmft Index were recorded in patients coming from Bangladesh, China, Romania, the Philippines, El Salvador, Albania, and Ukraine. Subjects coming from Sri Lanka had the lowest mean DMFT/dmft values. No correlation could be found between the prevalence of dental caries and annual sugar consumption per person; according to the latest research, sugar intakes in China, the Philippines, and Albania were equal to 6.5 kg, 21.1 kg, and 16.5 kg per capita in 2019, respectively, while the annual sugar intake in Sri Lanka was recorded to be equal to 27.6 kg in 2019 [42].

The oral health of children is influenced by the oral hygiene habits of the family members [43], since family represents the primary socializing agent for children, who consequently imitate their parents in their oral hygiene practices [44].

The literature reported that, aside from cultural habits, the oral health status of migrant children could be associated with the family's socioeconomic condition, parental occupation, parents' education level, and marital status [45]. Parental knowledge on oral health and oral-health related behaviour are crucial for the prevention of dental caries in children, and these mediating factors vary based on cultural and ethnic backgrounds [46]. It was also demonstrated that the social context of migrant families from middle- and low-human development countries may be unfavourable for children's oral health [46].

With regard to the CPI, significant statistical differences were found after adjusting data for the educational level of the included participants (p value < 0.01), but the results regarding the DMFT/dmft Index were not significant at p value < 0.01. To support our data, a recent study conducted in the USA demonstrated the association between severe periodontitis among Whites and African Americans and a low income and low education level [47].

According to several studies, poor oral health in the migrant population may be associated with different factors: low socioeconomic status, unemployment, low education, ethnic background, language difficulties, and inequalities in access to oral healthcare facilities.

5. Limitations of the Study

Although the number of subjects included in the study was significant, the majority of the sample was represented by children and adolescents, and only 12 adults out of 200 patients were enrolled. This could be considered as a limitation of our study, in particular, in the analysis of the periodontal tissue condition.

Our research included the patients who visited the dental department of the Policlinico Tor Vergata (Rome, Italy). These subjects visited the hospital dental centre, since they experienced dental problems, but it is assumed that there are many migrants without dental problems, who consequently do not need dental care. This point could represent a selection bias.

6. Conclusions

This observational study recorded poor oral health conditions among 200 migrants, from middle- and low-income countries, resident in Italy. It was demonstrated that the selected sample presented a high prevalence of dental caries, and a higher level of CPI was associated with a lower educational level.

Patient-oriented strategies and programmes are needed to improve the oral conditions of the migrant population.

Author Contributions: Conceptualization, data curation, writing—original draft preparation, writing—review and editing: D.L. and G.M.; investigation: M.M.; methodology: V.C.; supervision: C.A.; validation: F.C. All authors have read and agreed to the published version of the manuscript.

Funding: This research was funded by the Italian Health Ministry project titled "Dental caries prevention and screening among socially vulnerable subjects with difficult access to care". Codice CUP E89C2000020001.

Institutional Review Board Statement: This study was approved by the Fondazione Policlinico Tor Vergata (Rome, Italy) ethical committee (approval number N. 0001919/2020 del 29/01/2020).

Informed Consent Statement: Informed consent was obtained from all subjects involved in the study.

Data Availability Statement: Not applicable.

Conflicts of Interest: The authors declare no conflict of interest.

References

1. Batra, M.; Gupta, S.; Erbas, B. Oral health beliefs, attitudes, and practices of South Asian migrants: A systematic review. *Int. J. Environ. Res. Public Health* **2019**, *16*, 1952. [CrossRef]
2. Arora, G.; Mackay, D.F.; Conway, D.I.; Pell, J.P. Ethnic differences in oral health and use of dental services: Cross-sectional study using the 2009 Adult Dental Health Survey. *BMC Oral Health* **2016**, *17*, 1. [CrossRef] [PubMed]
3. IOM UN Migration. World Migration Report 2022. Available online: https://worldmigrationreport.iom.int/wmr-2022-interactive (accessed on 18 April 2022).
4. United Nations, Department of Economic and Social Affairs, Population Division. In *International Migration Report 2017: Highlights*; ST/ESA/SER.A/404; United Nations: New York, NY, USA, 2017.
5. Eurostat. Statistic Explained Migration and Migrant Population Statistics. Data Extracted in March 2022. Available online: https://ec.europa.eu (accessed on 16 April 2022).
6. Zucoloto, M.L.; Maroco, J.; Alvares Duarte Bonini Campos, J. Impact of oral health on health-related quality of life: A cross-sectional study. *BMC Oral Health* **2016**, *16*, 55. [CrossRef]
7. Sulaiman, L.; Saub, R.; Adinar Baharuddin, N.; Hasnur Safii, S.; Gopal Krishna, V.; Bartold, P.M.; Vaithilingam, R.D. Impact of Severe Chronic Periodontitis on Oral Health-related Quality of Life. *Oral Health Prev. Dent.* **2019**, *17*, 365–373. [CrossRef] [PubMed]
8. Kane, S.F. The effects of oral health on systemic health. *Gen. Dent.* **2017**, *65*, 30–34. [PubMed]
9. Gamonal, J.; Bravo, J.; Malheiros, Z.; Stewart, B.; Morales, A.; Cavalla, F.; Gomez, M. Periodontal disease, and its impact on general health in Latin America. Section I: Introduction part I. *Braz Oral Res.* **2020**, *34* (Suppl. S1), e024. [CrossRef]
10. Jin, L.J.; Lamster, I.B.; Greenspan, J.S.; Pitts, N.B.; Scully, C.; Warnakulasuriya, S. Global burden of oral diseases: Emerging concepts, management, and interplay with systemic health. *Oral Dis.* **2015**, *22*, 609–619. [CrossRef]
11. Oral Health. 2014. Available online: https://www.healthypeople.gov/2020/leading-health-indicators/2020-lhi-topics/Oral-Health (accessed on 15 April 2022).

12. Kandelman, D.; Arpin, S.; Baez, R.J.; Baehni, P.C.; Petersen, P.E. Oral health care systems in developing and developed countries. *Periodontology 2000* **2012**, *60*, 98–109. [CrossRef] [PubMed]
13. WHO. *The WHO Health Promotion Glossary*; WHO: Geneva, Switzerland, 1998.
14. Dao, T.H.; Docquier, F.; Maurel, M.; Schaus, P. Global migration in the twentieth and twenty-first centuries: The unstoppable force of demography. *Rev. World Econ.* **2021**, *157*, 417–449. [CrossRef]
15. Lauritano, D.; Moreo, G.; Carinci, F.; Campanella, V.; Della Vella, F.; Petruzzi, M. Oral Health Status among Migrants from Middle- and Low-Income Countries to Europe: A Systematic Review. *Int. J. Environ. Res. Public Health* **2021**, *18*, 12203. [CrossRef]
16. Riza, E.; Karnaki, P.; Gil-Salmerón, A.; Zota, K.; Ho, M.; Petropoulou, M.; Katsas, K.; Garcés-Ferrer, J.; Linos, A. Determinants of Refugee and Migrant Health Status in 10 European Countries: The Mig-HealthCare Project. *Int. J. Environ. Res. Public Health* **2020**, *17*, 6353. [CrossRef]
17. Lebano, A.; Hamed, S.; Bradby, H.; Gil-Salmerón, A.; Durá-Ferrandis, E.; Garcés-Ferrer, J.; Azzedine, F.; Riza, E.; Karnaki, P.; Zota, D.; et al. Migrants' and refugees' health status and healthcare in Europe: A scoping literature review. *BMC Public Health* **2020**, *20*, 1039. [CrossRef]
18. Scholten, P.; Entzinger, H.; Penninx, R. *Integrating Immigrants in Europe: Research-Policy Dialogues*; IMISCOE Research Series; Springer: Berlin/Heidelberg, Germany, 2015. [CrossRef]
19. Skeie, M.S.; Riordan, P.J.; Klock, K.S.; Espelid, I. Parental risk attitudes and caries-related behaviours among immigrant and western native children in Oslo. *Commun. Dent. Oral Epidemiol.* **2006**, *34*, 103–113. [CrossRef]
20. Dahlan, R.; Badri, P.; Saltaji, H.; Amin, M. Impact of acculturation on oral health among immigrants and ethnic minorities: A systematic review. *PLoS ONE* **2019**, *14*, e0212891. [CrossRef]
21. Gao, X.L.; McGrath, C. A review on the oral health impacts of acculturation. *J. Immigr. Minor. Health* **2011**, *13*, 202–213. [CrossRef]
22. Pabbla, A.; Duijster, D.; Grasveld, A.; Sekundo, C.; Agyemang, C.; van der Heijden, G. Oral Health Status, Oral Health Behaviours and Oral Health Care Utilisation Among Migrants Residing in Europe: A Systematic Review. *J. Immigr. Minor. Health* **2021**, *23*, 373–388. [CrossRef]
23. Serna, C.A.; Sanchez, J.; Arevalo, O.; Tomar, S.L.; McCoy, V.; Devieux, J.; De La Rosa, M.; Bastida, E. Self-reported factors associated with dental care utilization among Hispanic migrant farmworkers in South Florida. *J. Public Health Dent.* **2020**, *80*, 186–193. [CrossRef]
24. Svensson, I.; Gustafsson, J.; Uleskog, U.; Mathisson, C.; Mollai, N.; Kahlmeter, A.; Matsson, L. Oral condition and background factors in Somali immigrant children newly arrived in Sweden. *Swed. Dent. J.* **2016**, *40*, 153–164.
25. Reza, M.; Amin, M.S.; Sgro, A.; Abdelaziz, A.; Ito, D.; Main, P.; Azarpazhooh, A. Oral Health Status of Immigrant and Refugee Children in North America: A Scoping Review. *J. Can. Dent. Assoc.* **2016**, *82*, 1488–2159.
26. Brzoska, P.; Erdsiek, F.; Waury, D. Enabling and Predisposing Factors for the Utilization of Preventive Dental Health Care in Migrants and Non-Migrants in Germany. *Front. Public Health* **2017**, *5*, 201. [CrossRef]
27. Tiwari, T.; Albino, J. Acculturation and Pediatric Minority Oral Health Interventions. *Dent. Clin. N. Am.* **2017**, *61*, 549–563. [CrossRef] [PubMed]
28. Mouradian, W.E.; Huebener, C.E.; Ramos-Gomez, F.; Slavkin, H.C. Beyond access: The role of family and community in children's oral health. *J. Dent. Educ.* **2007**, *71*, 619–631. [CrossRef]
29. Moradi, G.; Mohamadi Bolbanabad, A.; Moinafshar, A.; Adabi, H.; Sharafi, M.; Zareie, B. Evaluation of Oral Health Status Based on the Decayed, Missing and Filled Teeth (DMFT) Index. *Iran. J. Public Health* **2019**, *48*, 2050–2057. [CrossRef] [PubMed]
30. Broadbent, J.M.; Thomson, W.M. For debate: Problems with the dmf index pertinent to dental caries data analysis. *Commun. Dent. Oral Epidemiol.* **2005**, *33*, 400–409. [CrossRef] [PubMed]
31. Roland, E.; Gueguen, G.; Longis, M.J.; Boiselle, J. Validation of the reproducibility of the dmf index used in bucco-dental epidemiology and evaluation of its 2 clinical forms. *World Health Stat. Q.* **1994**, *47*, 44–61.
32. Pitts, N.B.; Ekstrand, K.R. International Caries Detection and Assessment System (ICDAS) and its International Caries Classification and Management System (ICCMS)—Methods for staging of the caries process and enabling dentists to manage caries. *Commun. Dent. Oral Epidemiol.* **2013**, *41*, e41–e52. [CrossRef] [PubMed]
33. Nazir, M.; Al-Ansari, A.; Al-Khalifa, K.; Alhareky, M.; Gaffar, B.; Almas, K. Global Prevalence of Periodontal Disease and Lack of Its Surveillance. *Sci. World J.* **2020**, *2020*, 2146160. [CrossRef] [PubMed]
34. Benigeri, M.; Brodeur, J.M.; Payette, M.; Charbonneau, A.; Ismaïl, A.I. Community periodontal index of treatment needs and prevalence of periodontal conditions. *J. Clin. Periodontol.* **2000**, *27*, 308–312. [CrossRef]
35. Warnakulasuriya, S.; Kujan, O.; Aguirre-Urizar, J.M.; Bagan, J.V.; González-Moles, M.A.; Kerr, A.R.; Lodi, G.; Weber Mello, F.; Monteiro, L.; Ogden, G.R.; et al. Oral potentially malignant disorders: A consensus report from an international seminar on nomenclature and classification, convened by the WHO Collaborating Centre for Oral Cancer. *Oral Dis.* **2021**, *27*, 1862–1880. [CrossRef] [PubMed]
36. Muthukumar, S.; Suresh, R. Community periodontal index of treatment needs index: An indicator of anaerobic periodontal infection. *Indian J. Dent. Res.* **2009**, *20*, 423–425. [CrossRef]
37. Cutress, T.W.; Ainamo, J.; Sardo-Infirri, J. The community periodontal index of treatment needs (CPITN) procedure for population groups and individuals. *Int. Dent. J.* **1987**, *37*, 222–233.
38. Severino, M.; Caruso, S.; Ferrazzano, G.F.; Pisaneschi, A.; Fiasca, F.; Caruso, S.; De Giorgio, S. Prevalence of Early Childhood Caries (ECC) in a paediatric italian population: An epidemiological study. *Eur. J. Paediatr. Dent.* **2021**, *22*, 189–198. [CrossRef]

39. Campus, G.; Solinas, G.; Cagetti, M.G.; Senna, A.; Minelli, L.; Majori, S.; Montagna, M.T.; Reali, D.; Castiglia, P.; Strohmenger, L. National Pathfinder Survey of 12-Year-Old Children's Oral Health in Italy. *Caries Res.* **2007**, *41*, 512–517. [CrossRef]
40. Ferrazzano, G.F.; Cantile, T.; Sangianantoni, G.; Ingenito, A.; Rengo, S.; Alcidi, B.; Spagnuolo, G. Oral health status and Unmet Restorative Treatment Needs (UTN) in disadvantaged migrant and not migrant children in Italy. *Eur. J. Paediatr. Dent.* **2019**, *20*, 10–14. [CrossRef] [PubMed]
41. van Meijeren-van Lunteren, A.W.; Wolvius, E.B.; Raat, H.; Jaddoe, V.W.V.; Kragt, L. Ethnic background and children's oral health-related quality of life. *Qual. Life Res.* **2019**, *28*, 1783–1791. [CrossRef]
42. OurWorldinData. Available online: https://ourworldindata.org/grapher/daily-per-capita-caloric-supply (accessed on 27 May 2022).
43. Okada, M.; Kawamura, M.; Kaihara, Y.; Matsuzaki, Y.; Kuwahara, S.; Ishidori, H.; Miura, K. Influence of parents' oral health behaviour on oral health status of their school children: An exploratory study employing a causal modelling technique. *Int. J. Paediatr. Dent.* **2002**, *12*, 101–108. [CrossRef]
44. Mustafa, M.; Nasir, E.F.; Nordrehaug Åstrøm, A. Attitudes toward brushing children's teeth—A study among parents with immigrant status in Norway. *Int. J. Paediatr. Dent.* **2021**, *31*, 80–88. [CrossRef] [PubMed]
45. Ke, X.; Zhang, L.; Li, Z.; Tang, W. Inequality in health service utilization among migrant and local children: A cross-sectional survey of children aged 0–14 years in Shenzhen, China. *BMC Public Health* **2020**, *20*, 1668. [CrossRef]
46. Julihn, A.; Cunha Soares, F.; Hjern, A.; Dahllöf, G. Development level of the country of parental origin on dental caries in children of immigrant parents in Sweden. *Acta Paediatr.* **2021**, *110*, 2405–2414. [CrossRef]
47. Borrell, L.N.; Beck, J.D.; Heiss, G. Socioeconomic Disadvantage and Periodontal Disease: The Dental Atherosclerosis Risk in Communities Study. *Am. J. Public Health* **2006**, *96*, 332–339. [CrossRef] [PubMed]

applied sciences

MDPI

Article

Factors Affecting the Outcome of Periapical Surgery; a Prospective Longitudinal Clinical Study

Muhammad Adeel Ahmed [1,2,*], Nouman Mughal [3,4], Syed Hani Abidi [4], Muhammad Furqan Bari [5], Mohammed Mustafa [6], Fahim Vohra [7,8] and Ali Alrahlah [8,9]

1 Department of Operative Dentistry and Institute of Biomedical Sciences, Dow University of Health Sciences, Karachi 74200, Pakistan
2 Department of Restorative Dentistry and Endodontics, College of Dentistry, King Faisal University, Al-Ahsa 31982, Saudi Arabia
3 Department of Surgery, Aga Khan University, Karachi 74800, Pakistan; muhammad.nouman@aku.edu
4 Department of Biological and Biomedical Sciences, Aga Khan University, Karachi 74800, Pakistan; m.haniabidi@gmail.com
5 Department of Pathology, Dr. Ishrat-ul-Ebad Khan Institute of Oral Health Sciences, Dow University of Health Sciences, Karachi 74200, Pakistan; bari.m.f@gmail.com
6 Department of Conservative Dental Sciences, College of Dentistry, Prince Sattam Bin Abdulaziz University, Al-Kharj 11942, Saudi Arabia; ma.mustafa@psau.edu.sa
7 Department of Prosthetic Dental Science, College of Dentistry, King Saud University, Riyadh 11545, Saudi Arabia; fvohra@ksu.edu.sa
8 Engineer Abdullah Bugshan Research Chair for Dental and Oral Rehabilitation (Department), College of Dentistry, King Saud University, Riyadh 11545, Saudi Arabia; aalrahlah@ksu.edu.sa
9 Department of Restorative Dentistry—Operative Division, College of Dentistry, King Saud University, Riyadh 11545, Saudi Arabia
* Correspondence: mshakeel@kfu.edu.sa; Tel.: +96-65-8167-4914

Citation: Ahmed, M.A.; Mughal, N.; Abidi, S.H.; Bari, M.F.; Mustafa, M.; Vohra, F.; Alrahlah, A. Factors Affecting the Outcome of Periapical Surgery; a Prospective Longitudinal Clinical Study. *Appl. Sci.* **2021**, *11*, 11768. https://doi.org/10.3390/app112411768

Academic Editor: Gianrico Spagnuolo

Received: 23 August 2021
Accepted: 22 November 2021
Published: 10 December 2021

Abstract: Wound healing following periapical surgery is influenced by age, gender, smoking, periapical lesion size, type of root-end filling, method of root-end cavity preparation and the use of microsurgical or conventional technique. This study aimed to evaluate the influence of various preoperative factors such as age, gender, smoking, preoperative pain, and preoperative periapical lesion size on the outcome of surgical endodontic treatment. A thorough history, examination, and investigation were performed to establish patient age, gender, smoking status, periapical lesion size, and the presence of preoperative pain. Forty patients aged between 15–57 years presented with persistent chronic apical periodontitis of single-rooted anterior teeth after conventional re-root canal treatment were enrolled for periapical surgery. Following periapical surgery, all patients were recalled for evaluation of periapical healing after 12 months based on clinical and periapical X-ray examination due to inaccessibility of an advanced imaging system (CBCT). Chi-squared and Fisher's exact test were applied, which revealed a statistically significant association of periapical healing with age ($p = 0.025$), smoking ($p = 0.029$), and lesion size ($p < 0.001$). Although, the success of periapical healing was higher in males 78.6% (22/28) compared to females 58.3% (7/12) however, no statistically significant relationship was found between gender and healing ($p = 0.254$). Patient age, smoking status, and size of the preoperative lesion had a strong influence on periapical healing after surgical endodontic treatment.

Keywords: periapical surgery; surgical endodontic treatment; preoperative factors; prospective study

1. Introduction

The treatment of choice for symptomatic, persistent or enlarging periapical lesion after conventional re-root canal treatment is surgical endodontic treatment, which is aimed at resecting the root end and then creating a barrier at the apical end of the root with an inert material [1–3]. The goal of this treatment protocol is to salvage the tooth and prevent it from further damage by limiting the entry of microbes and their byproducts into the canal system, thus promoting the health of the tooth.

The success of surgical endodontic treatment mainly relies on the absence of clinical signs and symptoms and radiographic resolution of periapical lesions. Studies have shown that multiple factors may influence the outcome of periapical surgery such as age, gender, smoking, type of root-end filling material, size of the preoperative periapical lesion, and the presence of pretreatment signs and symptoms [3,4].

Smoking is considered one of the risk factors that can impede periapical healing. Smoking disturbs healing by limiting the supply of oxygenated blood and nutrients to the periapical area [5]. This in turn makes periapical tissue more susceptible to bacterial infection [6]. Additionally, it alters the synthesis of collagen, expedites bone loss, and interferes in the tissue repair process around the periapical area [7,8]. Furthermore, smokers have restricted defense mechanisms due to the deficiency of several immunological factors such as Tumor necrosis factor alpha (TNF-α) and Human beta defensin-2 (hBD-2) [9].

There have been numerous studies investigating the size of preoperative periapical lesions with periapical wound healing. It was observed that fibroblasts proliferate into a large bony defect after periapical surgery and form scar tissues instead of allowing osteoblasts for osseous regeneration [10], which in turn disrupts healing. However, the involvement of several molecular and cellular factors makes it more intricate [11]. Few studies have reported a better healing outcome with smaller lesions [3,12,13] while others find no statistically significant difference in outcome in relation to the size of periapical lesions [14,15].

The presence of preoperative signs and symptoms indicates the acute stage of periapical lesion. Acute lesions usually present with pain and swelling while chronic lesions are usually asymptomatic with or without the presence of sinus tract. A consensus has been observed that the presence of preoperative clinical signs and symptoms do not significantly influence postoperative healing [16,17] however, vice versa has also been reported in some studies [3,18].

In light of conflicting evidence and lack of prospective studies regarding the factors that determine the successful outcome, the current study was designed to evaluate the various preoperative factors affecting the outcome of surgical endodontic treatment.

2. Materials and Methods

2.1. Study Setting and Ethical Approval

This study was conducted at the Department of Operative Dentistry in collaboration with the Institute of Biomedical Sciences, Dow University of Health Sciences, Karachi from July 2017 to December 2018 after obtaining ethical approval (IRB-862/DUHS/Approval/2017/50) from Institutional Review Board, Dow University of Health Sciences. All the recommendations of the Helsinki Declaration and its later amendments were strictly followed.

2.2. Inclusion and Exclusion Criteria

Initially, 52 patients aged between 15–57 years who presented with chronic apical periodontitis of single-rooted anterior teeth, characterized radiographically as radiolucency greater than 4 mm around the root apex with loss of lamina dura and periodontal ligament space due to incompletely attempted or previously failed root canal treatment or periapical lesion around an open root apex in the tooth, where root formation was halted prematurely due to trauma, were enrolled in this study. Medically compromised patients with any uncontrolled systemic disease (ASA III), patients with pre-existing periodontal disease or patients in whom the lesion was smaller than 4 mm, multirooted teeth, teeth with cracks and root fractures, pregnant or lactating women and patients who did not turn up for follow-up visits were excluded from this study.

2.3. Clinical Procedure

History, examination, and investigation were performed to ascertain the presence of certain preoperative factors. Preoperative digital radiographs using the parallel technique were taken with a cone indicator and reference marker placed on the radiographic sensor Written informed consent was obtained from the patients.

Conventional re-root canal treatment: Local anesthesia 1:80,000 lidocaine with epinephrine was administered and a rubber dam was used for isolation. Access to the root was gained by removing the coronal restoration, followed by application of GP solvent (Carvene, PREVESTDenPro Prevest Denpro Limited, Jammu, India) and 25 or 30 # H file (Mani, Utsunomiya, Tochigi, Japan) for the removal of radicular filling. Working length was established by 35 or 40 # K file (Mani, Tochigi, Japan). Canal cleaning and shaping was performed by modified crown-down technique along with copious irrigation by 3.0% sodium hypochlorite (CanalPro NaOCl, Coltène/Whaledent, Altstätten, St. Gallen, Altstätten, Switzerland). Calcium hydroxide (Metapaste, Meta Biomed, Cheongju-si, Chungcheongbuk-do, Korea) was used as an intracanal medicament in between appointments. Finally, obturation was performed by cold lateral condensation technique followed by composite coronal restoration.

All patients were recalled after 3 to 6 months for clinical and radiographic evaluation of the periapical healing after conventional re-root canal treatment. Periapical healing was observed in seven patients and five patients failed to appear on follow-up visit, therefore 12 patients were excluded from the study. Finally, 40 patients underwent periapical surgery due to nonhealing periapical lesions (Figure 1).

Periapical Surgery: To begin with, local anesthesia 1:80,000 lidocaine with epinephrine was administered and a full thickness mucoperiosteal flap was elevated. After identifying the lesion site, access to the lesion was achieved by performing window preparation in cortical bone with small round bur No. 2 (Mani, Tochigi, Japan) in a slow-speed handpiece (Figure 2). The periapical lesion was identified and surgically removed by a surgical curette (Hibro Int, Tokyo, Japan). Zirconia-coated ultrasonic tip (Pro ultra, Maillefer DENTSPLY, Ballaigues, Switzerland) was used for apicoectomy and retrograde cavity preparation (Figure 3). The 3 mm apical fragment of the root was resected perpendicular to the long axis of the root with minimum or no bevel. Finally, MTA (Pro-root MTA, DENTSPLY Tulsa Dental Specialties, Johnson city, TX, USA) was used for retrograde filling (Figure 4). The flap was then repositioned and sutured with a 3/0 silk suture (ETHICON, Johnson & Johnsons, New Brunswick, NJ, USA). A periapical radiograph was taken after surgery. Analgesics and antibiotics were prescribed. These patients were recalled after 12 months for a follow-up visit. Both clinical and radiographic examination was performed at each follow-up visit (Figure 5). The parameters studied for clinical examination were pain, swelling, sinus tract, tenderness to palpation, tenderness to percussion, mobility, and discoloration. Radiographic healing was assessed using Periapical Index (PAI). The radiographic interpretation was carried out by two blinded endodontists. The interexaminer reliability in evaluating the postoperative radiographs was calculated and presented.

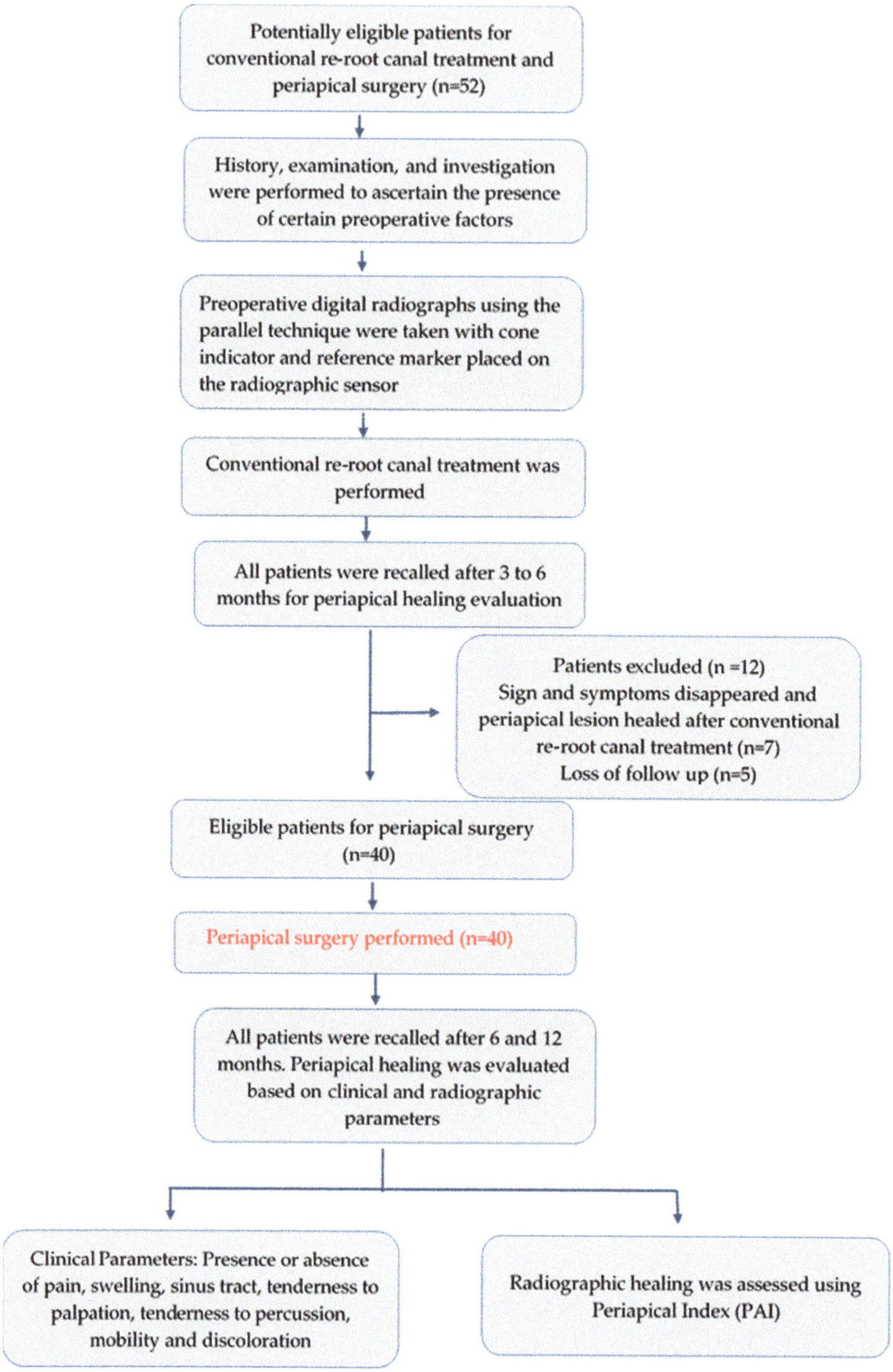

Figure 1. Flowchart of Methodology.

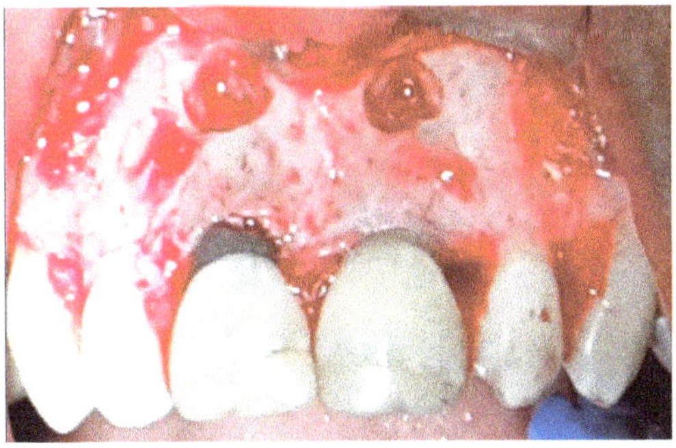

Figure 2. A full-thickness mucoperiosteal flap was reflected and the window was created around the periapical area of teeth 11 and 21.

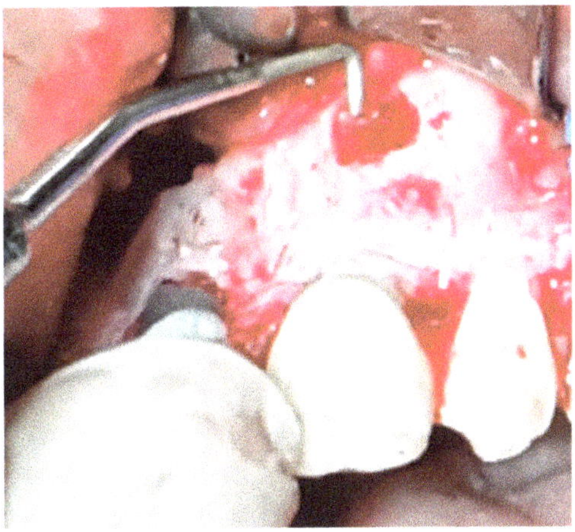

Figure 3. Retrograde cavity preparation was performed by zirconia-coated ultrasonic tip used for periapical surgery.

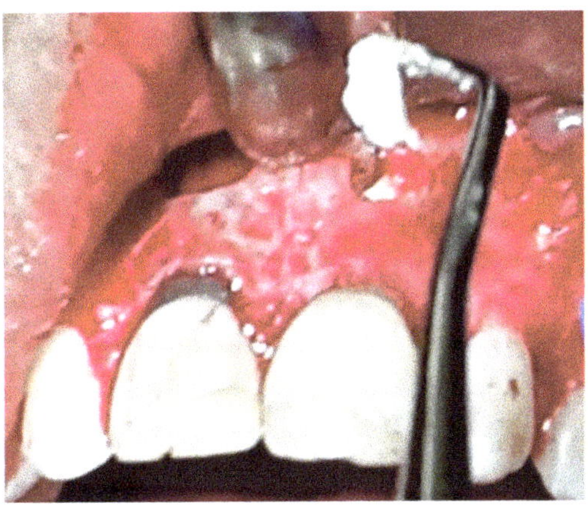

Figure 4. The retrograde filling was performed by MTA.

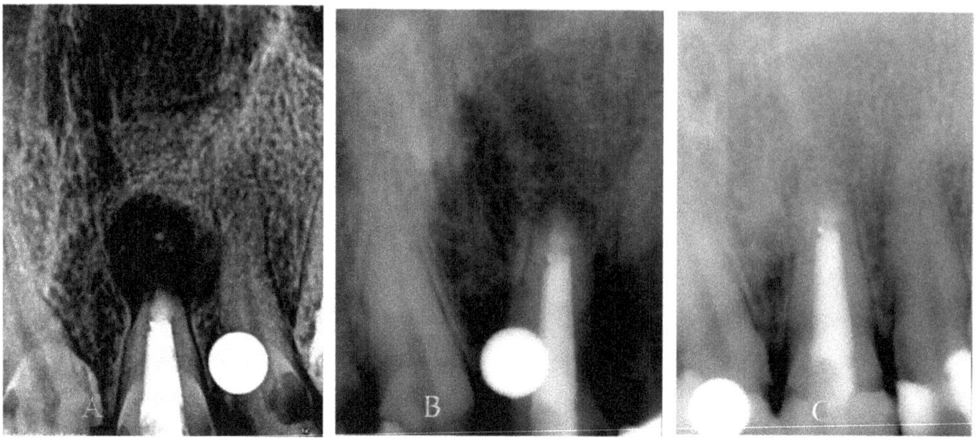

Figure 5. (**A**). Immediate postoperative periapical radiograph showing periapical radiolucency around the apex of Tooth 12. (**B**). Periapical radiolucency decreased in six months. (**C**). Complete resolution of periapical radiolucency in one year.

2.4. Statistical Analysis

The data were analyzed using a statistical package for social sciences (SPSS Version 25.0. Armonk, NY, USA). The results are presented as frequencies and percentages for age, gender, smoking, preoperative pain, size of the periapical lesion, and healing. Statistical associations were performed between periapical healing with age, gender, smoking, preoperative pain, size of the periapical lesion using chi-squared and Fisher's exact test as per data assumptions. A p-value of 0.05 or less was considered statistically significant.

3. Results

Out of 40 patients, 32 (80%) patients were below 30 years of age while only 8 (20%) patients were above the age of 30 years. The gender distribution showed that 88 (58.7%) of the patients were male and 62 (41.3%) were female. Most of the patients, 26 (65.0%), were nonsmokers. The size of the preoperative periapical lesion was arbitrarily categorized into 6 mm or less and more than 6 mm for this study. The results showed that half of the patients had a periapical lesion size of 6 mm or less, while another half had more than 6 mm. Preoperative pain was present in 22 patients (55.01%) (Table 1).

Table 1. Descriptive Statistics for Demographics.

Characteristics	N = 40 (%)
Age	
<30 years	32 (80.0)
>30 years	08 (20.0)
Gender	
Male	28 (70.0)
Female	12 (30.0)
Smoking	
Nonsmoker	26 (65.0)
Smoker	14 (35.0)
Preoperative pain	22 (55.0)
Periapical lesion size	
≤6 mm	20 (50.0)
>6 mm	20 (50.0)
Postoperative healing	29 (72.5)

There was a statistically significant association between periapical healing and age (p-value = 0.025), smoking (p-value = 0.029) and the size of periapical lesion (p-value < 0.001). After periapical surgery, healing was observed in 29 patients, in which 26 (81.3%) patients were below the age of 30 years while only 3 (37.5%) patients were above 30 years of age on one year follow-up. Similarly, 22 nonsmoking patients (84.6%) revealed periapical healing compared to 7 (50%) smoking patients. Although the success of healing was higher in males, at 78.6% (22/28), compared to females, at 58.3% (7/12), no statistically significant relationship was found between gender and healing (p-value = 0.254). Likewise, no significant association between the presence of preoperative pain and postoperative healing (p-value = 0.723) was observed (Table 2).

Table 2. Relationship of Periapical Healing with Age, Gender, Smoking, Preoperative pain, and Periapical lesion size on one year follow-up.

Characteristics	Periapical Healing			
	Yes (n = 29)	No (n = 11)	Total	p-Value
Age in years				
<30	26 (81.3%)	6 (18.8%)	32	0.025 ^
>30	3 (37.5%)	5 (62.5%)	8	
Gender				
Male	22 (78.6%)	6 (21.4%)	28	0.254 ^
Female	7 (58.3%)	5 (41.7%)	12	
Smoking				
Nonsmoker	22 (84.6%)	4 (15.4%)	26	0.029 ^
Smoker	7 (50.0%)	7 (50.0%)	14	
Preoperative pain				
Yes	15 (68.2%)	7 (31.8%)	22	0.723 ^
No	14 (77.8%)	4 (22.2%)	18	
Periapical lesion size				
≤6 mm	20 (100%)	0 (0.0%)	20	<0.001 ¥
>6 mm	09 (45.0%)	11 (55.0%)	20	

¥ Chi-squared test, ^ Fisher's Exact Test.

4. Discussion

The current study was conducted to evaluate the effect of various preoperative factors such as age, gender, smoking, preoperative pain, and preoperative periapical lesion size on the outcome of surgical endodontic treatment. The results of the study showed a statistically significant relationship between periapical wound healing and age, smoking, and preoperative lesion size.

Surgical endodontic retreatment is employed in managing recurrent secondary periapical lesions as a consequence of primary root canal treatment failure [19]. Surgical endodontic retreatment involves re-root canal treatment with subsequent sectioning of the apical end of the root followed by retrograde obturation and establishment of the apical seal of the root canal system. This apical seal facilitates healing of the periapical tissues by preventing the ingress of microbial irritants into the periapical area and concomitant tissue reaction [3].

It has now become clinically evident that even if the surgical endodontic treatment has been conducted using the same procedure, patients respond differently with respect to periapical wound healing [3,4]. This difference in periapical wound healing may be due to the presence of various preoperative factors such as pain, swelling, sinus tract, smoking, and size of the periapical lesion. However, evidence to support the influence of these preoperative factors on the outcome of periapical surgery are conflicting. In addition, most of the clinical studies that exist in the literature are retrospective [8,20,21]. The novelty of this study lies in following patients prospectively and opening an avenue for further research in this direction.

Age is considered a prognostic indicator of outcome for many surgical procedures [4]. In this study, 81.3% healing was achieved in patients below 30 years of age. This finding is in agreement with Kriesler et al. [22] They found a 95% success in patients between 21 to 40 years of age. Contrarily, Barone et al. [23] reported 84% success in periapical healing in patients above 45 years of age, compared to 68% in patients below 45 years of age. This study observed no significant relationship between healing and gender. Similarly, a large number of studies did not find gender as a prognostic factor for periapical surgery [18,24,25]. However, Peñarrocha-Diago et al. reported a higher success in males (60%) compared to females (40%) at 6 months [26].

The habit of smoking can influence the outcome of periapical surgery by several plausible biological mechanisms. First, smoking interferes with the function of lymphocytes, leukocytes, macrophages and increases the level of various proinflammatory mediators, such as TNF- α, IL-6, and C-reactive protein [27,28]. Second, smoking decreases the fibroblast migration to the periapical wound area and stimulates the function of osteoclastic cells, which encourage bone resorption [29]. Third, smoking causes morphological alteration in the microvasculature, which in turn disturbs the oxygen and nutrient supply to the periapical wound area [30]. In the current study, a positive trend of healing was observed in nonsmokers (84.6%). This finding conforms to Lopez et al. [8] and Kirkevang et al. [31], who found a statistically significant association between smoking and periapical healing. On the other hand, Rodriguez et al. [32] and Balto et al. [33] reported no significant difference in periapical periodontitis between smokers and nonsmokers.

The presence of preoperative pain and swelling or sinus tract may also govern the outcome of periapical surgery as indicated by many studies. Von Arc et al. [20] revealed that the presence of preoperative signs and symptoms discourages healing. They postulated that the healing potential of surgical wounds could be significantly affected by the stage of infection at the time of surgery, which is influenced by preoperative pain and signs. Similarly, Kreisler et al. [22] also claimed a lower success rate in patients with pretreatment pain. Contrastingly, the present study did not find any difference in the healing success rate in relation to the presence of preoperative pain, similarly to Song et al. [34] and Peñarrocha et al. [35]. The reason for there being no relation between the presence of preoperative pain and periapical healing after surgical endodontic treatment may be due to the fact that same treatment protocol was used for every patient and all treatment procedures were performed by the principal investigator.

Another important prognostic factor in periapical healing is the size of the lesion. In the present study, 100% successful healing was observed in patients with preoperative lesions of less than 6 mm, while 45% healing success was evident in patients with more than 6 mm of preoperative lesion size. A large number of studies have described the favorable prognosis after periapical surgery in patients with preoperative lesions of less than 5 mm in diameter [20,26,34]. Alternatively, Barone et al. [23] found an 80% success rate in periapical lesions of less than 10 mm in size at the time of surgery, compared to 53% success in periapical lesions of more than 10 mm in size.

Multiple systemic diseases are known to interfere with periapical healing, such as diabetes, hypertension, osteoporosis, or any uncontrolled systemic disease. Hyperglycemia increases the level of inflammatory markers and influences the various functions of the patient immune system [36]. Similarly, hypertension is associated with alterations in response and differentiation of bone cells at various levels [21]. Moreover, the drug bisphosphonate is commonly used for the treatment and prevention of osteoporosis and it is well-established that it reduces bone remodeling [37], therefore patients with these systemic diseases were excluded from this study, in addition to the patients with a pre-existing medical condition such as pregnancy and lactating mothers. Further prospective studies are required to evaluate postoperative healing in relation to the presence of different chronic systemic diseases using microsurgical techniques.

The findings of the study must be seen in light of some limitations. First, periapical surgery was performed by the conventional approach rather than microsurgical. The microsurgical approach offers numerous benefits such as small osteotomy, easy identification of root apices, visualization of minor anatomical variations, and preservation of cortical bone and root length by nearly 90-degree root resection. Several studies have shown the superiority of microsurgical endodontic treatment in comparison to traditional root-end surgery [38,39]. Second, the healing was assessed on a digital periapical radiograph, using a cone indicator and reference marker to ensure constant distance and angle between X-ray cone and sensor on every shoot. Moreover, exposure time, tube current, and voltage were the same on recall images. However, the X-ray image obtained was still two-dimensional, with a high probability of missing details in the third dimension. Nowadays, cone beam computed tomography (CBCT) is considered a standard of care [40] and has great value in establishing the correct diagnosis and in evaluating periapical healing after surgical endodontic treatment. Third, periapical surgery was performed on single-rooted anterior teeth only, multirooted posterior teeth where surgical access to treatment is difficult and high chances of variation in root morphology were excluded. Lastly, the sample size of the study was small due to the constrained study time, therefore; the results should be interpreted cautiously and may not represent the large population.

There is also a dire need to design a questionnaire which can evaluate the health-related quality of patients' life after endodontic surgery, similar to The University of Washington Quality of Life Questionnaire (UW-QOL) for head-and-neck cancer patients [41]. Future studies are also desirable with the inclusion of a large sample size, multirooted teeth and various other treatment-related factors such as microsurgical technique, type of root-end filling, method of root-end resection and retrograde cavity preparation. Addition of tools such as fractal analysis used for the quantitative evaluation of bone trabeculation following periapical surgery will also help in the early detection of complex structural patterns in the trabecular bone [42].

5. Conclusions

Patient age, smoking status, and size of the preoperative lesion had a strong influence on periapical healing after surgical endodontic treatment. However, other perioperative (treatment-related) factors should also be taken into account to establish conclusive evidence.

Author Contributions: Conceptualization, M.A.A., N.M. and S.H.A.; methodology, M.A.A.; software, M.A.A.; validation, F.V., M.M., A.A. and M.F.B.; formal analysis, M.A.A.; investigation, M.A.A. and A.A.; resources, N.M. and S.H.A.; data curation, M.F.B.; writing—original draft preparation, M.A.A. and A.A.; writing—review and editing, M.A.A., F.V., M.M., A.A. and S.H.A.; visualization, N.M.; supervision, N.M., S.H.A. and M.F.B.; project administration, M.A.A. and A.A.; funding acquisition, M.A.A. All authors have read and agreed to the published version of the manuscript.

Funding: The authors are grateful to the Deanship of Scientific Research, King Saud University for supporting through Vice Deanship of Scientific Research Chairs, Engr. Abdullah Bugshan research chair for Dental and Oral Rehabilitation.

Institutional Review Board Statement: The study was conducted according to the guidelines of the Declaration of Helsinki, and approved by the Institutional Review Board, Dow University of Health Sciences (IRB-862/DUHS/Approval/2017/50) on 10 July 2017.

Informed Consent Statement: Informed consent was obtained from all subjects involved in the study.

Data Availability Statement: The data presented in this study are available on request from the corresponding author. The data are not publicly available due to ethical concerns.

Conflicts of Interest: The authors declare no conflict of interest.

References

1. Schilder, H.; Hargreaves, K.M. Filling root canals in three dimensions. *J. Endod.* **2006**, *32*, 281–290. [CrossRef]
2. Alghamdi, F.; Alhaddad, A.J.; Abuzinadah, S. Healing of Periapical Lesions After Surgical Endodontic Retreatment: A Systematic Review. *Cureus* **2020**, *12*, e6916. [CrossRef]
3. Ahmed, M.A.; Anwar, M.F.; Ahmed, K.; Aftab, M.; Nazim, F.; Bari, M.F.; Mustafa, M.; Vohra, F.; Alrahlah, A.; Mughal, N.; et al. Baseline MMP expression in periapical granuloma and its relationship with periapical wound healing after surgical endodontic treatment. *BMC Oral Health* **2021**, *21*, 562. [CrossRef] [PubMed]
4. Serrano-Giménez, M.; Sánchez-Torres, A.; Gay-Escoda, C. Prognostic factors on periapical surgery: A systematic review. *Med. Oral Patol. Oral Cir. Bucal* **2015**, *20*, e715–e722. [CrossRef]
5. Segura-Egea, J.J.; Martín-González, J.; Castellanos-Cosano, L. Endodontic medicine: Connections between apical periodontitis and systemic diseases. *Int. Endod. J.* **2015**, *48*, 933–951. [CrossRef] [PubMed]
6. Krall, E.A.; Sosa, C.A.; Garcia, C.; Nunn, M.E.; Caplan, D.J.; Garcia, R.I. Cigarette smoking increases the risk of root canal treatment. *J. Dent. Res.* **2006**, *85*, 313–317. [CrossRef]
7. Peršić Bukmir, R.; Jurčević Grgić, M.; Brumini, G.; Spalj, S.; Pezelj-Ribaric, S.; Brekalo Pršo, I. Influence of tobacco smoking on dental periapical condition in a sample of Croatian adults. *Wien. Klin. Wochenschr.* **2016**, *128*, 260–265. [CrossRef]
8. López-López, J.; Jané-Salas, E.; Martín-González, J.; Castellanos-Cosano, L.; Llamas-Carreras, J.M.; Velasco-Ortega, E.; Segura-Egea, J.J. Tobacco smoking and radiographic periapical status: A retrospective case-control study. *J. Endod.* **2012**, *38*, 584–588. [CrossRef]
9. Ghattas Ayoub, C.; Aminoshariae, A.; Bakkar, M.; Ghosh, S.; Bonfield, T.; Demko, C.; Montagnese, T.A.; Mickel, A.K. Comparison of IL-1β, TNF-α, hBD-2, and hBD-3 Expression in the Dental Pulp of Smokers Versus Nonsmokers. *J. Endod.* **2017**, *43*, 2009–2013. [CrossRef]
10. Melcher, A.H. On the Repair Potential of Periodontal Tissues. *J. Periodontol.* **1976**, *47*, 256–260. [CrossRef]
11. Lin, L.; Chen, M.Y.H.; Ricucci, D.; Rosenberg, P.A. Guided Tissue Regeneration in Periapical Surgery. *J. Endod.* **2010**, *36*, 618–625. [CrossRef] [PubMed]
12. Von Arx, T.; Jensen, S.S.; Hänni, S.; Friedman, S. Five-year longitudinal assessment of the prognosis of apical microsurgery. *J. Endod.* **2012**, *38*, 570–579. [CrossRef] [PubMed]
13. Çalişkan, M.K.; Tekin, U.; Kaval, M.E.; Solmaz, M.C. The outcome of apical microsurgery using MTA as the root-end filling material: 2- to 6-year follow-up study. *Int. Endod. J.* **2016**, *49*, 245–254. [CrossRef] [PubMed]
14. Shinbori, N.; Grama, A.M.; Patel, Y.; Woodmansey, K.; He, J. Clinical Outcome of Endodontic Microsurgery That Uses EndoSequence BC Root Repair Material as the Root-end Filling Material. *J. Endod.* **2015**, *41*, 607–612. [CrossRef]
15. Lui, J.N.; Khin, M.M.; Krishnaswamy, G.; Chen, N.N. Prognostic factors relating to the outcome of endodontic microsurgery. *J. Endod.* **2014**, *40*, 1071–1076. [CrossRef]
16. Lustmann, J.; Friedman, S.; Shaharabany, V. Relation of pre- and intraoperative factors to prognosis of posterior apical surgery. *J. Endod.* **1991**, *17*, 239–241. [CrossRef]
17. Allen, R.K.; Newton, C.W.; Brown, C.E. A statistical analysis of surgical and nonsurgical endodontic retreatment cases. *J. Endod.* **1989**, *15*, 261–266. [CrossRef]
18. Rahbaran, S.; Gilthorpe, M.S.; Harrison, S.D.; Gulabivala, K. Comparison of clinical outcome of periapical surgery in endodontic and oral surgery units of a teaching dental hospital: A retrospective study. *Oral Surg. Oral Med. Oral Pathol. Oral Radiol. Endodontology* **2001**, *91*, 700–709. [CrossRef]
19. Kruse, C.; Spin-Neto, R.; Christiansen, R.; Wenzel, A.; Kirkevang, L.L. Periapical Bone Healing after Apicectomy with and without Retrograde Root Filling with Mineral Trioxide Aggregate: A 6-year Follow-up of a Randomized Controlled Trial. *J. Endod.* **2016**, *42*, 533–537. [CrossRef]
20. Von Arx, T.; Peñarrocha, M.; Jensen, S. Prognostic factors in apical surgery with root-end filling: A meta-analysis. *J. Endod.* **2010**, *36*, 957–973. [CrossRef]
21. Martins, C.M.; Sasaki, H.; Hirai, K.; Andrada, A.C.; Gomes-Filho, J.E. Relationship between hypertension and periapical lesion: An in vitro and in vivo study. *Braz. Oral Res.* **2016**, *30*, e78. [CrossRef] [PubMed]
22. Kreisler, M.; Gockel, R.; Aubell-Falkenberg, S.; Kreisler, T.; Welte, C.; Filippi, A.; Kühl, S.; Schütz, S.; D'Hoedt, B. Clinical outcome in periradicular surgery: Effect of patient- and tooth-related factors—A multicenter study. *Quintessence Int.* **2013**, *44*, 53–60.
23. Barone, C.; Dao, T.T.; Basrani, B.B.; Wang, N.; Friedman, S. Treatment Outcome in Endodontics: The Toronto Study-Phases 3, 4, and 5: Apical Surgery. *J. Endod.* **2010**, *36*, 28–35. [CrossRef]
24. Zuolo, M.L.; Ferreira, M.O.F.; Gutmann, J.L. Prognosis in periradicular surgery: A clinical prospective study. *Int. Endod. J.* **2000**, *33*, 91–98. [CrossRef]
25. Wesson, C.M.; Gale, T.M. Molar apicectomy with amalgam root-end filling: Results of a prospective study in two district general hospitals. *Br. Dent. J.* **2003**, *195*, 707–714. [CrossRef] [PubMed]
26. Peñarrocha-Diago, M.; Ortega-Sánchez, B.; García-Mira, B.; Maestre-Ferrín, L.; Peñarrocha-Oltra, D.; Gay-Escoda, C. A prospective clinical study of polycarboxylate cement in periapical surgery. *Med. Oral Patol. Oral Cir. Bucal* **2012**, *17*, 276–280. [CrossRef] [PubMed]
27. Johnson, G.K.; Guthmiller, J.M. The impact of cigarette smoking on periodontal disease and treatment. *Periodontol. 2000* **2007**, *44*, 178–194. [CrossRef]

28. Cabanillas-balsera, D.; Segura-Egea, J.J.; Bermudo-fuenmayor, M.; Martín-gonzález, J.; Jiménez-sánchez, M.C.; Areal-quecuty, V.; Sánchez-domínguez, B.; Montero-miralles, P.; Velasco-Ortega, E. Smoking and Radiolucent Periapical Lesions in Root Filled Teeth: Systematic Review and Meta-Analysis. *J. Clin. Med.* **2020**, *9*, 3506. [CrossRef]
29. Bergström, J.; Babcan, J.; Eliasson, S. Tobacco smoking and dental periapical condition. *Eur. J. Oral Sci.* **2004**, *112*, 115–120.
30. Krall, E.A.; Sosa, C.A.; Garcia, C.; Nunn, M.E.; Caplan, D.J.; Garcia, R.I. Cigarette smoking is associated with an acute impairment of microvascular function in humans. *Clin. Sci.* **2003**, *104*, 247–252.
31. Kirkevang, L.L.; Wenzel, A. Risk indicators for apical periodontitis. *Community Dent. Oral Epidemiol.* **2003**, *31*, 59–67. [CrossRef]
32. Rodriguez, F.R.; Taner, B.; Weiger, R.; Walter, C. Is smoking a predictor of apical periodontitis? *Clin. Oral Investig.* **2013**, *17*, 1947–1955. [CrossRef]
33. Balto, H.A.; Alabdulaaly, L.; Bahammam, S.; Al-Ekrish, A.A. Comparative analysis of prevalence of apical periodontitis in smokers and non-smokers using cone-beam computed tomography. *Saudi Dent. J.* **2019**, *31*, 52–57. [CrossRef] [PubMed]
34. Song, M.; Jung, I.Y.; Lee, S.J.; Lee, C.Y.; Kim, E. Prognostic factors for clinical outcomes in endodontic microsurgery: A retrospective study. *J. Endod.* **2011**, *37*, 927–933. [CrossRef] [PubMed]
35. Peñarrocha Diago, M.; Ortega Sánchez, B.; García Mira, B.; Martí Bowen, E.; Von Arx, T.; Gay Escoda, C. Evaluation of healing criteria for success after periapical surgery. *Med. Oral Patol. Oral Cir. Bucal* **2008**, *13*, 143–147.
36. Segura-Egea, J.J.; Martín-González, J.; Cabanillas-Balsera, D.; Fouad, A.F.; Velasco-Ortega, E.; López-López, J. Association between diabetes and the prevalence of radiolucent periapical lesions in root-filled teeth: Systematic review and meta-analysis. *Clin. Oral Investig.* **2016**, *20*, 1133–1141. [CrossRef]
37. Hsiao, A.; Glickman, G.; He, J. A retrospective clinical and radiographic study on healing of periradicular lesions in patients taking oral bisphosphonates. *J. Endod.* **2009**, *35*, 1525–1528. [CrossRef] [PubMed]
38. Tortorici, S.; Difalco, P.; Caradonna, L.; Tetè, S. Traditional endodontic surgery versus modern technique: A 5-year controlled clinical trial. *J. Craniofac. Surg.* **2014**, *25*, 804–807. [CrossRef]
39. Setzer, F.C.; Shah, S.B.; Kohli, M.R.; Karabucak, B.; Kim, S. Outcome of endodontic surgery: A meta-analysis of the literature—part 1: Comparison of traditional root-end surgery and endodontic microsurgery. *J. Endod.* **2010**, *36*, 1757–1765. [CrossRef]
40. Abdinian, M.; Katiraei, M.; Zahedi, H.; Rengo, C.; Soltani, P.; Spagnuolo, G. Age Estimation Based on Pulp–Tooth Volume Ratio of Anterior Teeth in Cone-Beam Computed Tomographic Images in a Selected Population: A Cross-Sectional Study. *Appl. Sci.* **2021**, *11*, 9984. [CrossRef]
41. Lowe, D.; Rogers, S. University of Washington Quality of Life Questionnaire (UW-QOL v4 and v4.1). Available online: http://www.headandneckcancer.co.uk/sites/default/files/assets/pages/UW-QOL-update-2018.pdf (accessed on 10 June 2021).
42. Soltani, P.; Sami, S.; Yaghini, J.; Golkar, E.; Riccitiello, F.; Spagnuolo, G. Application of Fractal Analysis in Detecting Trabecular Bone Changes in Periapical Radiograph of Patients with Periodontitis. *Int. J. Dent.* **2021**, *7*, 3221448. [CrossRef] [PubMed]

applied sciences

MDPI

Systematic Review

Oral Manifestations in Scurvy Pediatric Patients: A Systematic Review and a Case Report

Maria Francesca Gicchino [1], Antonio Romano [2], Speranza Cioffi [1], Fausto Fiori [2], Emanuele Miraglia del Giudice [1], Alberta Lucchese [2,*], Alma Nunzia Olivieri [1] and Rosario Serpico [2]

1 Department of Woman and Child and General and Specialized Surgery, University of Campania Luigi Vanvitelli, Via De Crecchio 6, 80138 Naples, Italy; mariafrancesca.gicchino@unicampania.it (M.F.G.); speranza.cioffi@unicampania.it (S.C.); emanuele.miragliadelgiudice@unicampania.it (E.M.d.G.); almanunzia.olivieri@unicampania.it (A.N.O.)

2 Multidisciplinary Department of Medical-Surgical and Dental Specialties, University of Campania Luigi Vanvitelli, Via De Crecchio 6, 80138 Naples, Italy; antonio.romano4@unicampania.it (A.R.); fausto.fiori@outlook.com (F.F.); rosario.serpico@unicampania.it (R.S.)

* Correspondence: alberta.lucchese@unicampania.it

Abstract: Scurvy is generated by lack of vitamin C; although it is considered a rare and past disease, scurvy continues to be detected in children with neurodevelopmental disorders and with selective diet habits. Identifying scurvy can be demanding due to the perceived rarity of the condition, and it can become a tricky diagnostic question given to the variety of nonspecific symptoms, including gingival manifestations. This study aims to identify most common clinical features in order to provide a complete picture of the signs and symptoms, and to offer clinicians the diagnostic tools for identifying patients suffering from scurvy. We present a case report of a child affected by scurvy; it has also been performed as a systematic review about scurvy in pediatric population. A search yielded 107 relevant studies since 1990. Most of the identified cases have shown oral, musculoskeletal and cutaneous manifestation that improved within a few days of starting vitamin C therapy. Identifying scurvy's characteristic clinical features allows a timely diagnosis, thus avoiding invasive investigations. Pediatric dentists should possess adequate knowledge and experience to identify the main characteristics of scurvy. This can help facilitate a prompt diagnosis in order to provide timely intervention to the patient that is relatively ease and safe.

Keywords: clinical manifestations; gingival disorders; oral scurvy; scurvy; vitamin C deficiency

Citation: Gicchino, M.F.; Romano, A.; Cioffi, S.; Fiori, F.; Miraglia del Giudice, E.; Lucchese, A.; Olivieri, A.N.; Serpico, R. Oral Manifestations in Scurvy Pediatric Patients: A Systematic Review and a Case Report. *Appl. Sci.* **2021**, *11*, 8323. https://doi.org/10.3390/app11188323

Academic Editor: Bruno Chrcanovic

Received: 26 July 2021
Accepted: 6 September 2021
Published: 8 September 2021

1. Introduction

Ascorbic acid (vitamin C) plays the key role of a cofactor in several metabolic reactions involved in tissue growth, development and healing [1]. In fact, ascorbic acid enables collagen's hydroxylation [2]; then, it empowers the biosynthesis of carnitine and norepinephrine, and is also involved in the metabolism of tyrosine and the amidation of peptide hormones [3]. Vitamin C performs an important function in the immune system too: scorbutic individuals usually develop a higher incidence of infections, and vitamin C auxiliary therapy is actually considered for sepsis treatment support [4]. Generally, signs of ascorbic acid deficiency begin to show after about 30 to 90 days of insufficient vitamin C intake [5]; clinical findings are directly related to the various metabolic pathway which ascorbic acid is involved in [6]. Most of characteristic disorders of vitamin C deficiency can overlap with rheumatological, infectious or hematological diseases, showing a wide range of musculoskeletal and mucocutaneous manifestations, thus mimicking other pediatric conditions [7]. Scurvy is a well-known but uncommon disease, and nowadays is considered a rare condition in developed nations [8]. Despite its low frequency in the population, cases of scurvy still occur in people at risk, including elderly populations, patients affected by malabsorption syndromes and eating disorders, and, above all, pediatric patients with restricted or selective feeding [9]. These categories of people have chronically low levels of vitamin C, which cannot be intrinsically produced within the human body and therefore must be obtained via dietary intake [10]. The most common clinical features are hypertrophy, swelling and bleeding of the gums, follicular hyperkeratosis, lower limbs swelling and tenderness, poor wound healing [11,12]. Oral cavity signs represent peculiar characteristics in patients affected by scurvy; nevertheless, gingival overgrowth stands out as a typical manifestation of several diseases, resulting in differential diagnosis for idiopathic enlargement, drug-induced enlargement, enlargement associated with systemic diseases (such as Leukemia or Granulomatous diseases) and neoplastic enlargement (gingival tumors) [13,14]. Serological measurement of vitamin C is a readily available and widely used laboratory test; despite generally ascorbic acid serum levels accounts for the most recent food intake, they may be related to the level of reserves and predict the possibility of developing clinical signs [15]. Dietary restrictions and poor compliance with taking oral supplements appears to have brought scurvy back to the fore in the pediatric population, particularly in patients with neurodevelopmental disorders linked to highly selective diets [16]. Despite scurvy's return to clinical practice, the extreme heterogeneity of the clinical signs makes the identification of patients suffering from scurvy a very demanding challenge for the pediatric dentist [17]. The purpose of this study is to provide pediatric dentists with the means to identify clinical parameters and diagnostic tools useful for detecting patients suffering from scurvy. Here we present a case report of a patient with scurvy and a systematic review of the literature on scurvy in pediatric population.

2. Case Report

A four-year-old child presented to our department with lower limbs pain and refusal to walk, with legs fixed in flexion at hips and knees for about four months. The symptoms were partially managed with non-steroidal anti-inflammatory drugs (NSAIDs). Patient's mother did not refer fever, skin rash or weight loss. About familial history, his sister was healthy, his grandmother suffering from celiac disease and a first-grade cousin was affected from juvenile idiopathic arthritis (JIA) in therapy with NSAIDs. Patient was born at term with normal weight and length. His psychomotor development was referred in the norm up to seven–eight months of age, then patient presented a delay of psychomotor and language development. Dietary history reveals highly selective eating since the second year, based exclusively on ham and white meat homogenized, with refusal of fruits and vegetables. Since about one year the mother reports difficulty in eating, and several episodes of gingivitis with antivirals and topical antifungals with partial benefit. When patient came to our department he was in fair general conditions, his weight was 17 kg (25–50th) and his height was 105 cm (25–50th). The physical examination was difficult because of the patient's developmental delay, however it revealed pale and dry skin, corkscrew hair and signs of follicular hyperkeratosis in the lower and upper limbs. Child refused to walk, with legs fixed in flexion at hips and knees ("Frog leg-position"). Both legs were diffusely tender to palpation. The child was uncooperative for oral examination but erythematous, hemorrhagic, and swollen gums in maxillary anterior region were noted (Figure 1). The rest of physical examination was normal. Blood examinations revealed low iron and vitamin D level, while complete blood count, serum chemistries, liver and kidney, function panel, coagulation panel (prothrombin time, activated partial thromboplastin time, and international normalized ratio) and inflammatory parameters (C-reactive protein and erythrocyte sedimentation rate) were normal. Hips, knees and ankle ultrasound did not reveal joints effusion or signs of tenosynovitis. X-ray of the lower limbs (Figure 2) did not reveal fractures, but generalized osteopenia and typical features of malnutrition, including a ground glass appearance, Pelkan spur, which represents a healing metaphyseal pathologic fracture, and a Wimberger ring sign, which denotes a thin sclerotic cortex surrounding a lucent epiphysis. Periosteal reaction due to subperiosteal hemorrhage with a dense provisional calcification immediately adjacent to the physis (Frankel line), and an adjacent lucent band more diaphyseal in location (Trummerfeld line) were noted. In consideration of the result of blood examinations and of the lower limbs X-ray, we hypothesized a state of nutritional deficiency. Furthermore, peculiar elements of both personal history and clinical examination, such as selective dietary habits, follicular hyperkeratosis, corkscrew hair, relapsing gingivitis and joint pain in the absence of signs of local inflammation led to the diagnostic hypothesis of vitamin C deficiency. This clinical suspicion was confirmed by the finding of low levels serum vitamin C < 2.4 μmol/L (normal value 26.1–84.6). Supplementary treatment with oral vitamin C (300 mg daily) and D (800 UI daily) was started. Patient clinical condition improved with recovery of walking and an oral clinical healing of the gingival enlargement (Figure 3). One month after discharge, the boy had normal vitamins' levels: 35.6 μmol/l (normal value 26.1–84.6).

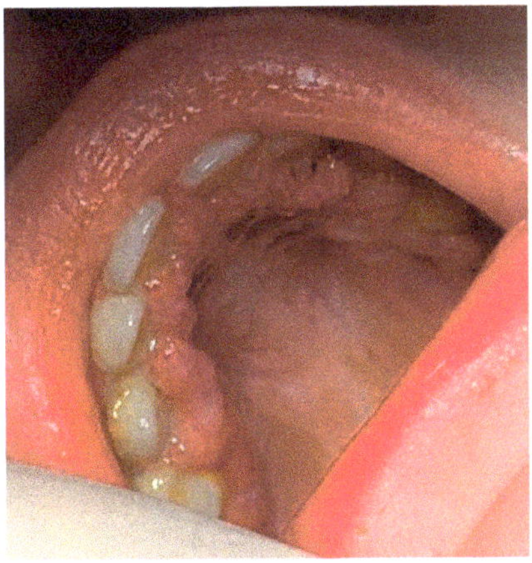

Figure 1. First oral clinical presentation (T0).

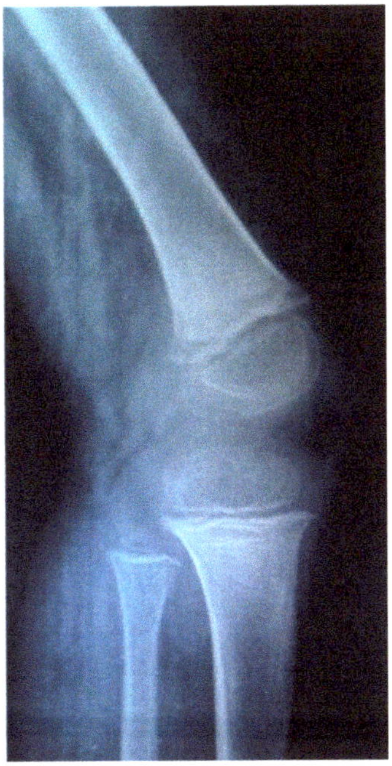

Figure 2. Lower limb X-ray showing typical features such as radio-dense band (Frenkel line), marginal spur formation.

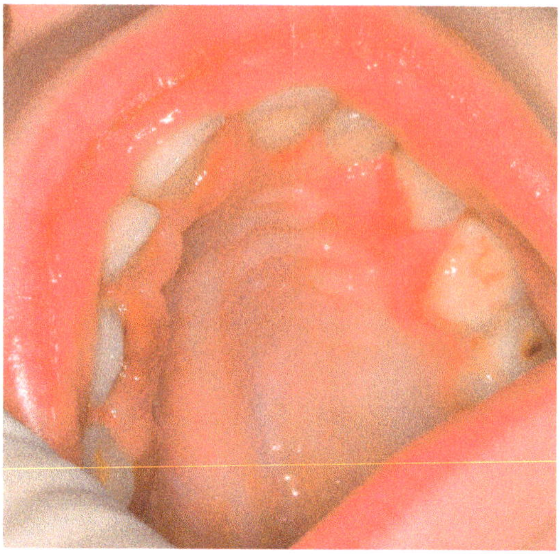

Figure 3. Oral tissues healing presentation after four weeks of Vitamin C administration (Tf).

3. Materials and Methods

A systematic review of the literature was performed through PubMed, ISI Web of science, and Cochrane Library. "Scurvy", "vitamin C deficiency", "Moeller's disease", "Cheadle's disease", "scorbutus", "Barlow's disease", "hypoascorbemia", "lack of vitamin C", "scorbutic", "child", "children", "pediatric", "toddler", "infant", "infancy" and "childhood" were employed as Medical Subject Headings (MeSH) terms (Table 1). Search operations have been completed in December 2020. The PICOS criteria and PRISMA checklist have been fulfilled in the review execution [18,19]. Male and female children (age $\leq$ 16 y), with confirmed diagnosis of scurvy, characterized by any kind of clinical manifestation related to vitamin C deficiency was set as population of interest (intervention); "no intervention" was comparison. Comparative studies, cross-sectional studies, retrospective studies, prospective studies, survey studies, case series and case reports were included. We aimed to identify the clinical manifestations and diagnostic methods of scurvy. The inclusion criteria of the selected studies were: presence of any clinical manifestation in humans, pediatric age, edited in English language, published since 1990. Review articles and studies without full text available were excluded. Three reviewers, once the initial results were collected, analyzed the titles and abstracts; then duplicates were excluded, and all those articles that did not match the inclusion criteria were ruled out. The full texts of the remaining articles were read in depth by two reviewers to better assess the content of the studies: demographic data, clinical manifestations, diagnostic path, therapeutic approach and all the extracted data have been organized in Table S1. The Risk of Bias in Non-randomized Studies of Interventions (ROBINS I) assessment tool was employed to evaluate quality of non-randomized studies [20]. This tool analyzes seven bias domains and each one refers the Risk of Bias (RoB) in five grades: low (LR), moderate (MR), serious (SR), critical (CR) and no information. The overall evaluation is based on the combination of these seven domains [21] (Table S2). A study based on a non-randomized design rarely presents a low level of RoB. The review was submitted and registered on PROSPERO [22] (registration code: CRD42021225174).

Table 1. The entire list used in the search and the combinations used in the research phase.

Search Topic								
	and #10			and #10			and #10	
	and #11			and #11			and #11	
	and #12			and #12			and #12	
#1	and #13	OR	#2	and #13	OR	#3	and #13	OR
	and #14			and #14			and #14	
	and #15			and #15			and #15	
	and #16			and #16			and #16	
	and #10			and #10			and #10	
	and #11			and #11			and #11	
	and #12			and #12			and #12	
#4	and #13	OR	#5	and #13	OR	#6	and #13	OR
	and #14			and #14			and #14	
	and #15			and #15			and #15	
	and #16			and #16			and #16	
	and #10			and #10			and #10	
	and #11			and #11			and #11	
	and #12			and #12			and #12	
#7	and #13	OR	#8	and #13	OR	#9	and #13	
	and #14			and #14			and #14	
	and #15			and #15			and #15	
	and #16			and #16			and #16	

1. Scurvy; 2. Vitamin C deficiency; 3. Moeller's disease; 4. Cheadle's disease; 5. Scorbutus; 6. Barlow's disease; 7. Hypoascorbemia; 8. Lack of vitamin C; 9. Scorbutic; 10. Child; 11. Children; 12. Childhood; 13. Infant; 14. Infancy; 15. Toddler; 16. Pediatric.

4. Results

Systematic Review of the Literature

The initial search yielded 719 results; 171 articles were excluded as they were duplicates. Another 373 articles were excluded following the reading of titles and abstracts. Reading the full text of the 175 remaining articles, 68 were excluded because they did not meet our inclusion criteria. At the end of the selection, 107 articles were included in this systematic review (Figure 4). Details on the 107 selected items and their main contents such as number of cases, clinical information, diagnostic path, and therapeutic protocol are summarized in Supplementary Table S1. Through the evaluation of the 107 analyzed studies, 88 are classified as overall moderate RoB (MR), and 19 as serious overall RoB (SR). Definitely, in the analysis of 107 non-randomized studies, no article appears to have a critical RoB in the individual domain or in the analysis of the overall domain and, therefore, all the studies confirm evidence (Supplementary Table S2). A total number of 134 patients were described within the 107 selected studies, with a minimum age of 6 months old and a maximum age of 16 years (mean 6.06 $\pm$ 3.81). Only 31 out of the 134 patients (23.1%) were female, and 103 (76.9%) were male. A total of 69 (51.5%) patients were affected by neurodevelopmental disorders, and 41 (31.3%) suffered of autistic spectrum disorder. On the clinical examination, 66 out of 134 (49.3%) patients showed cutaneous manifestations of scurvy, 92 of 134 (68.7%) patients showed musculoskeletal manifestation and 112 out of 134 (83.6%) patients showed oral manifestation. Among these 112 patients with oral manifestations, 1 of 112 (0.89%) was affected by oral ulcers, 1 of 112 (0.89%) had a chronic glossitis and 110 of 112 (98.21%) had gingival involvement, such as gingival bleeding, swelling and hypertrophy. The diagnosis was based on clinical hypothesis for 132 of the 134 patients (99.3%); serological dosage was obtained in 103 out of 134 patients (76.9%), histological examination was performed in 27 patients (20.1%); an ex juvantibus diagnostic approach was carried out in 17 cases (12.7%). The mean vitamin C serum level was 0.105 ± 0.118 mg/dl ranging from 0.000 to 0.700 mg/dl. Therapy was described in 78/134 patients. Only one patient of the 78 (1.28%) did not undergo vitamin C administration, but was only prescribed dietary recommendations. The remaining 77 patients (98.72%) underwent vitamin C therapy with a mean dosage of 384.81 ± 272.74 mg ranging from 50 to 1500 mg daily.

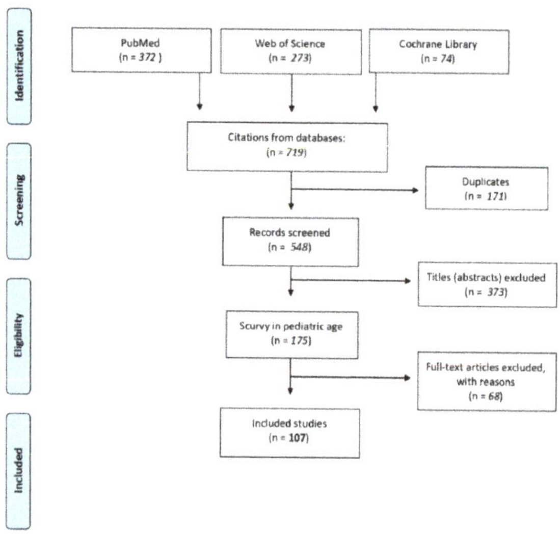

Figure 4. PRISMA flowchart.

5. Discussion

Main findings encountered in our case report, in fact, meet with the manifestations most commonly reported in the literature. Hahn et al. described the data obtained from 77 patients but did not mention any oral manifestation [12]. A recent literature review by Kothari et al. yielded 77 cases and gingival swelling, pain and bleeding were noted in all but three of the cases in which intraoral findings were described [23]. A review by Ratanachu-Ek et al. reported 86% of pediatric scurvy cases were initially misdiagnosed; a correct diagnostic approach should be based on the recognition of clinical manifestations [24]. Oral soft tissue involvement was the most striking sign that we observed in our patient, and it was also described in 92 out of the 107 selected studies (83.7% of examined patients). This important observation is the reason why clinicians that treat children and pediatric dentists should consider atypical gingival swelling or other gingival manifestations as one of the early signs of scurvy [25]. Scurvy is rare, but it still occurs among children with autism and developmental disorders, so this condition should be kept in mind in a clinical constellation of gingival involvement, lower extremity pain, limp, non-blanching rash, fatigue, anemia, in particular in children with an history of selective diet [26]. Having a comprehensive dietary history as part of the data gathering is fundamental to the early recognition of nutritional deficiency diseases in order to avoid invasive procedures and/or their severe complications [27].

6. Conclusions

The heterogeneity and time delay of diagnostic path encountered in the analyzed studies reveal a gap in our knowledge: a more focused clinical approach and a more detailed investigation on diet habits, neurodevelopmental disorders and on dermatological or orthopedic status should make up for this knowledge gap, in order to put dentists among the first actors against scurvy suspicions. This will lead to:

1. early detection and reduction in a misdiagnosis of scurvy
2. early intervention of vitamin C deficiency
3. subsequent prevention of the morbidities of this extremely distressing condition in pediatric patients.

Supplementary Materials: The following are available online at https://www.mdpi.com/article/10.3390/app11188323/s1. Table S1 Data extraction from included items. Table S2 Risk of Bias assessment.

Author Contributions: R.S. and E.M.d.G. conceived the ideas; M.F.G. and A.R. collected the data; Material A.L. and A.N.O. led the writing-review; F.F. and S.C. analyzed the data; A.L., A.N.O. and M.F.G. developed the methodology; R.S. and E.M.d.G. ensured validation and supervision. All authors have read and agreed to the published version of the manuscript.

Funding: This research received no external funding.

Institutional Review Board Statement: The study was conducted according to the guidelines of the Declaration of Helsinki.

Informed Consent Statement: Written informed consent has been obtained from the patient to publish this paper.

Data Availability Statement: Data shared are in accordance with consent provided by participants on the use of confidential data.

Conflicts of Interest: The authors declare no conflict of interest.

References

1. Padayatty, S.J.; Levine, M. Vitamin C: The Known and the Unknown and Goldilocks. *Oral Dis.* **2016**, *22*, 463–493. [CrossRef] [PubMed]
2. Pullar, J.M.; Carr, A.C.; Vissers, M.C.M. The Roles of Vitamin C in Skin Health. *Nutrients* **2017**, *9*, 866. [CrossRef]
3. Corpe, C.P.; Eck, P.; Wang, J.; Al-Hasani, H.; Levine, M. Intestinal Dehydroascorbic Acid (DHA) Transport Mediated by the Facilitative Sugar Transporters, GLUT2 and GLUT8. *J. Biol. Chem.* **2013**, *288*, 9092–9101. [CrossRef] [PubMed]
4. Fujii, T.; Luethi, N.; Young, P.J.; Frei, D.R.; Eastwood, G.M.; French, C.J.; Deane, A.M.; Shehabi, Y.; Hajjar, L.A.; Oliveira, G.; et al. Effect of Vitamin C, Hydrocortisone, and Thiamine vs Hydrocortisone Alone on Time Alive and Free of Vasopressor Support among Patients with Septic Shock: The VITAMINS Randomized Clinical Trial. *J. Am. Med. Assoc.* **2020**, *323*, 423–431. [CrossRef] [PubMed]
5. He, H.; Qiao, Y.; Zhang, Z.; Wu, Z.; Liu, D.; Liao, Z.; Yin, D.; He, M. Dual Action of Vitamin C in Iron Supplement Therapeutics for Iron Deficiency Anemia: Prevention of Liver Damage Induced by Iron Overload. *Food Funct.* **2018**, *9*, 5390–5401. [CrossRef]
6. Smirnoff, N. Ascorbic Acid Metabolism and Functions: A Comparison of Plants and Mammals. In *Free Radical Biology and Medicine*; Elsevier Inc.: Amsterdam, The Netherlands, 2018; pp. 116–129. [CrossRef]
7. Perkins, A.; Sontheimer, C.; Otjen, J.P.; Shenoi, S. Scurvy Masquerading as Juvenile Idiopathic Arthritis or Vasculitis with Elevated Inflammatory Markers: A Case Series. *J. Pediatr.* **2020**, *218*, 234–237.e2. [CrossRef] [PubMed]
8. Fortenberry, M.; Rucker, H.; Gaines, K. Pediatric Scurvy: How an Old Disease Is Becoming a New Problem. *J. Pediatr. Pharmacol. Ther.* **2020**, *25*, 735–741. [CrossRef]
9. Agarwal, A.; Shaharyar, A.; Kumar, A.; Bhat, M.S.; Mishra, M. Scurvy in Pediatric Age Group—A Disease Often Forgotten? *J. Clin. Orthop. Trauma* **2015**, *6*, 101–107. [CrossRef]
10. Byard, R.W.; Maxwell-Stewart, H. Scurvy-Characteristic Features and Forensic Issues. *Am. J. Forensic Med. Pathol.* **2019**, *40*, 43–46. [CrossRef]
11. Basu, T.K.; Donaldson, D. SCURVY. In *Encyclopedia of Food Sciences and Nutrition*; Elsevier: Amsterdam, The Netherlands, 2003; pp. 5107–5111. [CrossRef]
12. Hahn, T.; Adams, W.; Williams, K. Is Vitamin C Enough? A Case Report of Scurvy in a Five-Year-Old Girl and Review of the Literature. *BMC Pediatr.* **2019**, *19*, 1–6. [CrossRef]
13. Sharp, W.G.; Berry, R.C.; Burrell, L.; Scahill, L.; McElhanon, B.O. Scurvy as a Sequela of Avoidant-Restrictive Food Intake Disorder in Autism: A Systematic Review. *J. Dev. Behav. Pediatr.* **2020**, *41*, 397–405. [CrossRef] [PubMed]
14. Chapple, I.L.C.; Mealey, B.L.; Van Dyke, T.E.; Bartold, P.M.; Dommisch, H.; Eickholz, P.; Geisinger, M.L.; Genco, R.J.; Glogauer, M.; Goldstein, M.; et al. Periodontal Health and Gingival Diseases and Conditions on an Intact and a Reduced Periodontium: Consensus Report of Workgroup 1 of the 2017 World Workshop on the Classification of Periodontal and Peri-Implant Diseases and Conditions. *J. Periodontol.* **2018**, *89*, S74–S84. [CrossRef]

15. Robitaille, L.; Hoffer, L.J. A Simple Method for Plasma Total Vitamin C Analysis Suitable for Routine Clinical Laboratory Use. *Nutr. J.* **2016**, *15*, 1–9. [CrossRef]
16. Vaezipour, N.; Leibundgut, K. Nonalimental Scurvy with Relapse Symptoms after Stopping Oral Vitamin C Supplementation. *Pediatrics* **2018**, *142*, e20172139. [CrossRef]
17. Willmott, N.S.; Bryan, R.A. Case Report: Scurvy in an Epileptic Child on a Ketogenic Diet with Oral Complications. *Eur. Arch. Paediatr. Dent.* **2008**, *9*, 148–152. [CrossRef] [PubMed]
18. Saaiq, M.; Ashraf, B. Modifying "Pico" Question into "Picos" Model for More Robust and Reproducible Presentation of the Methodology Employed in A Scientific Study. *World J. Plast. Surg.* **2017**, *6*, 390–392.
19. Shamseer, L.; Moher, D.; Clarke, M.; Ghersi, D.; Liberati, A.; Petticrew, M.; Shekelle, P.; Stewart, L.A.; Altman, D.G.; Booth, A.; et al. Preferred Reporting Items for Systematic Review and Meta-Analysis Protocols (Prisma-p) 2015: Elaboration and Explanation. *BMJ* **2015**, *349*, g7647. [CrossRef]
20. Hinneburg, I. ROBINS-1: A Tool for Asssessing Risk of Bias in Non-Randomised Studies of Interventions. Available online: https://pubmed.ncbi.nlm.nih.gov/29952172/ (accessed on 20 April 2021).
21. Romano, A.; Di Stasio, D.; Lauritano, D.; Lajolo, C.; Fiori, F.; Gentile, E.; Lucchese, A. Topical Photodynamic Therapy in the Treatment of Benign Oral Mucosal Lesions: A Systematic Review. *J. Oral Pathol. Med.* **2020**, *50*, 639–648. [CrossRef]
22. Page, M.J.; Shamseer, L.; Tricco, A.C. Registration of Systematic Reviews in PROSPERO: 30,000 Records and Counting. *Syst. Rev.* **2018**, *7*, 1–9. [CrossRef]
23. Kothari, P.; Tate, A.; Adewumi, A.; Kinlin, L.M.; Ritwik, P. The Risk for Scurvy in Children with Neurodevelopmental Disorders. *Spec. Care Dent.* **2020**, *40*, 251–259. [CrossRef]
24. Ratanachu-Ek, S.; Jeerathanyasakun, Y.; Sukswai, P.; Wongtapradit, L. Scurvy in Pediatric Patients: A Review of 28 Cases. *J. Med. Asso. Thail.* **2003**, *86*, S734–S740.
25. Lund, R.M.; Becker, M.L.; Shapiro, S.; Allison, T.; Harris, J.G. Scurvy Presenting with Limp and Weakness: A Case Report. *BMC Pediatr.* **2019**, *19*, 228. [CrossRef] [PubMed]
26. Ma, N.S.; Thompson, C.; Weston, S. Brief Report: Scurvy as a Manifestation of Food Selectivity in Children with Autism. *J. Autism Dev. Disord.* **2016**, *46*, 1464–1470. [CrossRef] [PubMed]
27. Brennan, C.M.; Atkins, K.A.; Druzgal, C.H.; Gaskin, C.M. Magnetic Resonance Imaging Appearance of Scurvy with Gelatinous Bone Marrow Transformation. *Skelet. Radiol.* **2012**, *41*, 357–360. [CrossRef] [PubMed]

applied sciences

MDPI

Article

Comparison of Marginal Integrity and Surface Roughness of Selective Laser Melting, CAD-CAM and Digital Light Processing Manufactured Co-Cr Alloy Copings

Abdulaziz S. Alqahtani [1], Abdullah M. AlFadda [2], Malek Eldesouky [2], Mazen K. Alnuwaiser [2], Samar Al-Saleh [1], Saad Alresayes [1], Abdullah Alshahrani [1], Fahim Vohra [1] and Tariq Abduljabbar [1,*]

1 Department of Prosthetic Dental Science, College of Dentistry, King Saud University, Riyadh 11545, Saudi Arabia; absalqahtani@ksu.edu.sa (A.S.A.); Salsaleh@ksu.edu.sa (S.A.-S.); salresayes@ksu.edu.sa (S.A.); asalshahrani@ksu.edu.sa (A.A.); fvohra@ksu.edu.sa (F.V.)

2 Department of General Dentistry, College of Dentistry, King Saud University, Riyadh 11545, Saudi Arabia; alfadda35@gmail.com (A.M.A.); aldasouki.malik@gmail.com (M.E.); mazenalnuwaiser@gmail.com (M.K.A.)

* Correspondence: tajabbar@ksu.edu.sa; Tel.: +966-013477444

Abstract: The purpose of the present study was to evaluate the influence of fabrication techniques on the surface micro-roughness (Ra) and marginal misfit of cobalt chromium (CoCr) copings. A mandibular first molar was prepared for a metal ceramic crown. Forty metal copings were prepared and divided into groups (n = 10). Group 1, Casting-Lost wax technique (Cast-LWT), Group 2, CAD-CAM, Group 3, Selective laser melting (SLM), and Group 4, Digital light processing-Cast (DLP-Cast). Ra was measured using laser profilometry and marginal misfit was analyzed with Micro-CT. Analysis of variance (ANOVA), Tukey multiple comparison, and correlation coefficient tests were applied ($p < 0.05$). SLM technique showed the highest Ra (2.251 ± 0.310 μm) and the Cast-LWT group presented the lowest Ra (1.055 ± 0.184 μm). CAD-CAM copings showed statistically lower Ra compared with SLM samples ($p = 0.028$), but comparable Ra to DLP-Cast ($p > 0.05$). CoCr copings fabricated from the DLP-Cast technique demonstrated the highest marginal misfit (147.746 ± 30.306 μm) and the lowest misfit was established by SLM copings (27.193 ± 8.519 μm). The SLM technique displayed lower marginal misfit than DLP-Cast and CAD-CAM ($p = 0.001$), but comparable misfit to Cast-LWT copings. Ra influenced the marginal misfit in CAD-CAM, SLM, and DLP-Cast technique-fabricated copings. ($p < 0.01$). Marginal misfit and Ra of CoCr copings are contingent on the different fabrication techniques.

Keywords: marginal misfit; laser profilometry; Micro-CT; selective laser melting; CAD-CAM

Citation: Alqahtani, A.S.; AlFadda, A.M.; Eldesouky, M.; Alnuwaiser, M.K.; Al-Saleh, S.; Alresayes, S.; Alshahrani, A.; Vohra, F.; Abduljabbar, T. Comparison of Marginal Integrity and Surface Roughness of Selective Laser Melting, CAD-CAM and Digital Light Processing Manufactured Co-Cr Alloy Copings. *Appl. Sci.* **2021**, *11*, 8328. https://doi.org/10.3390/app11188328

Academic Editor: Gianrico Spagnuolo

Received: 28 July 2021
Accepted: 4 September 2021
Published: 8 September 2021

1. Introduction

Porcelain-fused-to-metal (PFM) crowns have been widely used in dentistry as a fixed dental aesthetic replacement for decades [1]. Increased aesthetic demands, advancements in casting techniques, availability of different alloys, and their use in almost all clinical conditions account for the popularity of PFM crowns [2]. Precise marginal fit of dental cast restoration is considered as the most critical and technical feature for the long-term successful clinical outcomes. Available evidence advocates an acceptable range of marginal misfit of full veneer crowns to be 100 to 120 μm [3]. However, multiple studies suggest different ranges of acceptable marginal misfit (10 to 160 μm) [4]. Increased marginal discrepancy accounts for the 10% of prosthetic failures, i.e., exposure of dental cement to the oral environment, bacterial penetration and plaque retention, secondary carious lesions, negative pulp reactions, marginal discoloration, periodontal disturbances, and esthetic and functional compromise [3].

Primarily, marginal integrity of cast copings are influenced by the surface characteristics, cast adaptation, and luting adhesives used [5]. Similarly, surface micro-roughness

(Ra) also influences the retention of indirect restoration [3]. It has been proposed that different fabrication techniques exhibit different Ra levels for cast indirect restorations. The casting with lost-wax technique (Cast-LWT) is commonly used to fabricate metal copings for PFM crowns [6,7]. The process involves the carving of inlay wax on a dye that mimics the anatomy and morphology of the lost tooth followed by spruing, investing, burnout, and casting with metal ingots [8]. However, it is time consuming and technique and operator sensitive to achieve a better quality of casting restoration [9]. In order to overcome these shortcomings, alternate contemporary digitally advanced fabrication techniques have been developed.

Among various techniques, computer-aided design/manufacturing (CAD-CAM) provides better standardization [10]. This method works on the principle of subtractive manufacturing, involving the milling of a solid metal block in to a desired shape of indirect prosthesis [11]. CAD-CAM offers standard-quality restorations, time efficiency, dimensional accuracy, and reduced risk of health hazards [12,13]. However, subtractive technique is associated with material wastage along with increased cost and difficult accessibility [14]. Current literature demonstrates conflicting outcomes related to the effect of Ra on the marginal misfit of metal copings fabricated using the subtractive technique [15,16].

Limitations related to the complexity of design obtained by subtractive technique led to introduction of novel additive fabrication techniques including selective laser melting (SLM), selective laser-sintering (SLS), and digital light processing (DLP) [17]. SLM is also referred to as 3D printing which is based on addition of material layer by layer using high power-density laser to melt and fuse non-precious alloy powders, particularly titanium and cobalt-chromium [18]. It can produce complex restoration designs without wasting a large amount of material. It is also able to produce multiple parts at the same time. However, it is an expensive technology and requires digital learning with operator skills [19]. Moreover, Digital light projection (DLP) is a further development in 3D Printing [20]. DLP printers project a silhouette of an entire layer simultaneously and polymerize it with a single shot of curing light with faster printing [21]. This method is rapid, precise, and affordable compared with the SLM technique along with providing a better finish [22]. However, data related to the influence of these fabrication methods on the Ra and misfit of metal copings need further investigation.

In a recent study by Kim et al., misfit of restorations was compared among CAD-CAM, Cast, and SLM methods. It was reported that SLM specimens showed higher misfit compared with CAD-CAM and cast crowns [23]. In addition, the current literature is limited on evidence to determine a gold standard among fabrication techniques of metal restorations [24,25]. It was hypothesized that there is no difference in Ra and marginal misfit of CoCr copings manufactured from conventional Cast-LWT and contemporary techniques (CAD-CAM, SLM, DLP-Cast). Therefore, the purpose of the present study was to evaluate the effect of different fabrication techniques on the Ra and marginal misfit of CoCr copings.

2. Materials and Methods

The study compared marginal misfit and surface micro-roughness of metal copings fabricated using Cast-LWT, CAD-CAM, SLM, and DLP-Cast.

2.1. Specimen Preparation

An ivorine mandibular first molar (KaVo Dental, Fruehauf Drive, Charlotte, NC, USA) was prepared for metal ceramic crown with a high-speed airotor and diamond burs (NSK Co., Japan). Tooth reduction parameters included 2 mm occlusal and axial, two-plane reduction on buccal surface, a radial shoulder margin, and a taper of 6° using a milling machine. Preparation margins were finished with a chisel and smoothed with silicon impregnated burs (Rubberrized abrasives, Lasco Diamond Products, Chatsworth, CA, USA) (Figure 1). The prepared tooth surface was recorded using polyvinyl siloxane (PVS) impression material (Putty light technique) and the impression was verified. A replica was prepared in wax and was casted using the lost wax technique in nickel-chrome alloy (Remanium, Dentauram GmbH & Co., Ispringen, Germany).

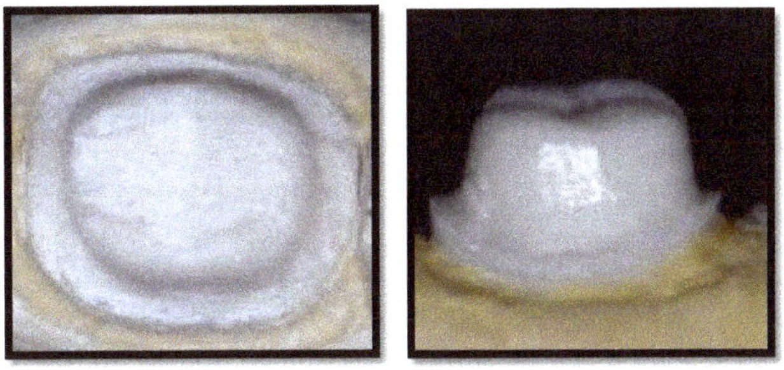

(A) (B)

Figure 1. Representative image of the prepared tooth specimen. Prepared tooth (**A**) Occlusal, (**B**) Buccal.

Forty metal alloy copings were prepared in the study and divided into study groups according to the fabrication technique (n = 10): Casting-Lost wax technique (Cast-LWT) (control), CAD-CAM, Selective laser melting (SLM), and Digital light processing with casting (DLP CAST). The sample size in each group of the study was calculated by performing a power calculation utilizing data from a similar previous study [11].

2.2. Casting-Lost Wax Technique (Cast LWT)

Wax copings with 0.7 mm thickness and 0.2 mm marginal reinforcements were prepared to fabricate cast CoCr copings by a senior dental technician. The wax patterns were sprued and invested using phosphate-bonded investment (Fast Fire 15 investment; Whip Mix, Louisville, KY, USA) with a 16 mL/60 g liquid/powder ratio with a ring-less technique. A Whip Mix plastic ring was used with bars to provide expansion, and the molds were removed from the ring after 15 min. After removal of investment, the molds were allowed to set for 24 h. Wax pattern burn-out was performed using a furnace (PROGRAMIX 50, Ugin'Dentaire, Seyssinet-Pariset, France) at 900 °C. CoCr alloy casting was carried out using a casting machine (FORNAX 35E®, BEGO, Bremen, Germany) at 1500 °C temperature (Wirebond ®C; BEGO, Bremen, Germany) (composition Co 63.3% Cr 24.8% W 5.3% Mo 5.1% Si 1.0%). The copings were divested with glass beads (50 µm) at 1 bar pressure, followed by ultrasonic cleaning.

2.3. Computer Aided Design-Computer Aided Manufacture (CAD-CAM)

Master die surface was coated with a uniform layer of Cercon Eye Scan Spray (DeguDent GmbH, 63457 Hanau-Wolfgang, Germany) and scanning was performed using Cercon Eye scanner (DeguDent GmbH, 63457 Hanau-Wolfgang, Germany). The scan was run using Cercon Art and contours were mapped and the final image in a steriolithographic (STL) format was displayed. The copings were designed using Cercon Art software according to the prescribed dimensions. Cercon Brain (DeguDent GmbH, Hanau-Wolfgang, Germany) milling machine fabricated the copings in the prescribed design. Ceramill Sintron alloy blanks (Co-Cr-Amann Girrbach AG, Herrschaftswiese, Koblach, Austria) were secured in the milling machine and were removed on milling completion.

2.4. Selective Laser Melting (SLM)

To fabricate SLM CoCr alloy copings, the STL file for coping design fabricated for the CAD-CAM technique was transferred to the Concept Laser Machine (metal laser melting system; GE Additive company, Boston, MA, USA) with standard parameters. CoCr alloy (Starbond Easy Powder 30; Scheftner GmbH, Mainz, Germany) (composition, Co

61%, Cr 27.5%, W 8.5%, Si 1.6%, C, Fe and Mn < 1%) with an alloy powder grain size of +10/−30 μm and elastic modulus of 225 GPa was used. The coping model was vertically positioned, and the support material for the printing was designed and attached within the design software. The printing process was carried out in nitrogen and argon inter atmosphere. The fiber laser beam (100 W ytterbium (Yb)) hit the powder layer in selective areas and created a melt pool resulting in the fusion of powder particles. The thickness of the powder layer was 20 μm. This process was repeated until the coping fabrication was completed.

2.5. Digital Light Processing-Cast (DLP Cast)

The fourth group was the three-dimensional-printed resin patterns using digital light processing (DLP) (M-One; MAKEX Technology, Zhejiang, China). A 3D printer (MiiCraft 125; MiiCraft, Jena, Germany) was used with a photo-polymerized biocompatible polymer resin (Freeprint Temp; DETAX GmbH & Co., Ettlingen, Germany). The printer settings included 50 μm thickness, 405 nm wavelength, and a curing time of 2.40 s per layer. The resin copings were casted to metal copings and measured before the final adjustments. A similar casting process was employed as described in earlier Section 2.2.

All coping samples were assessed for surface micro roughness and marginal misfit on the master die replica.

2.6. Assessment of Surface Micro Roughness (Ra)

The average surface Ra was calculated in micrometers (μm) using 3D optical non-contact laser profilometry (LPM) (Contour GT-K 3D Optical Microscope, Bruker®, Tucson, AZ, USA). The scanning parameters included magnification of 5× with a lens with a single window of 1 mm × 1 mm, 1× scan speed, and 3% thresholding. The copings were fixed horizontally on the stage using a mold fabricated with impression material (polyvinyl siloxane, 3M ESPE, St. Paul, MN, USA). The laser beam of the optical microscope was placed on the coping surface with the help of stage movement to obtain a good-quality image. The copings were scanned at five location points and an average was identified. To manage the precision and surface roughness parameters, a Vision 64 Control and Analysis Software (Bruker®, Tucson, AZ, USA) was used.

2.7. Marginal Misfit

The coping misfit was assessed in micrometer (μm). Misfit was analyzed with Bruker micro CT (Skyscan 1173 high-energy spiral scan micro-CT; Skyscan NV, Kontich, Belgium) to detect micro gaps at selected points. The coping samples were mounted and positioned inside the specimen chamber and the parameters included 130 kV of source energy at 60 μA at 300 ms of exposure time. To create a good image, a brass filter of 0.25 mm with a 0.2° rotation step for a 360° angle and 4-frame average was employed. Post scanning, reconstruction of 3D images was performed using N-Recon® software (program version 1.6.1.3, Bruker Skyscan, Kontich, Belgium). During this process, parameter adjustments were performed to enhance image quality. Reconstructed images were loaded in the Dataviewer® Software (Bruker Skyscan, Kontich, Belgium) to determine image quality and perform image misfit assessments. The measurements were performed between the marginal surface of the coping and the prepared marginal surface of replicas of master die (Dental stone).

All assessed recordings for surface roughness and marginal misfit were logged in an Excel sheet and mean and standard deviations were evaluated. Data were analyzed using Analysis of variance (ANOVA) and Tukey Kramer multiple comparisons test. Correlation between specimen roughness and misfit was assessed using Pearson correlation. A p value of <0.05 was considered statistically significant among groups.

3. Results

3.1. Surface Roughness

Mean and standard deviation for Ra scores of CoCr specimens after using different fabrication techniques are presented in Table 1. SLM fabricated copings displayed the highest mean values (2.251 ± 0.310 µm). Cast/LWT specimens presented the lowest mean Ra score (1.055 ± 0.184 µm). Ra values among all investigated groups were statistically significant ($p < 0.05$). SLM specimens showed significantly higher Ra than DLP cast copings (1.590 ± 0.167 µm) ($p = 0.001$). Mean Ra of Cast/LWT copings was lower than CAD-CAM (1.840 ± 0.236 µm) ($p = 0.001$), SLM ($p = 0.001$), and DLP cast ($p < 0.05$) specimens respectively. The micrographs obtained for surface roughness assessment are presented in Figure 2. The uniform high roughness was observed in SLM samples (2 D) and the DLP-Cast specimen displayed a smoother surface with minimal localized craters (2 C).

Table 1. Comparison of surface micro-roughness of the CoCr copings.

Study Group	Mean (µm)	SD	ANOVA *p* Value	Cast/LWT	CAD-CAM	SLM	DLP-Cast
Cast/LWT	1.055	0.184		1.000			
CAD-CAM	1.840	0.236	0.001 $	0.001 *	1.000		
SLM	2.251	0.310		0.001 *	0.038 *	1.000	
DLP-Cast	1.590	0.167		0.028	0.057	0.001 *	1.000

* Statistical significant difference using Tukey Kramer post hoc test; $ Statistical significant difference using ANOVA.

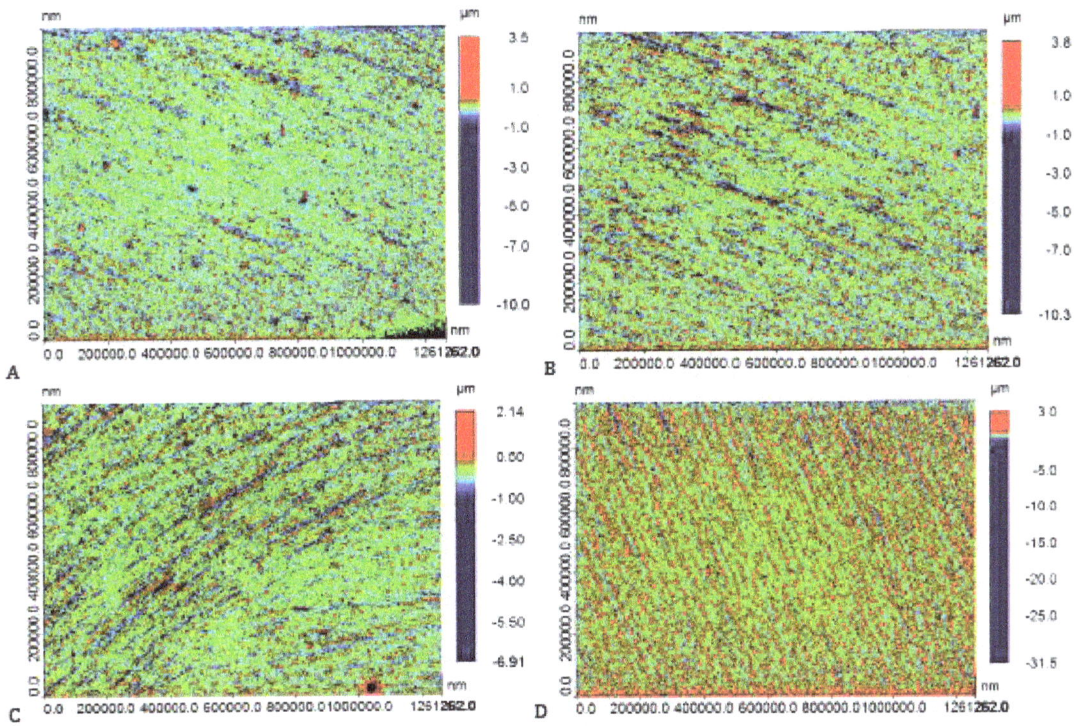

Figure 2. Surface roughness (Ra) micrographs for groups (**A**) Cast/LWT, (**B**) CAD-CAM, (**C**) DLP-Cast, and (**D**) SLM.

3.2. Marginal Misfit

Means and standard deviation of marginal misfit of CoCr specimens are presented in Table 2. Samples fabricated from the DLP-Cast technique demonstrated the highest mean marginal misfit (147.746 ± 30.306 μm), whereas the lowest marginal misfit was established by SLM-fabricated specimens (27.193 ± 8.519 μm). Moreover, ANOVA revealed that there was a statistically significant difference in mean marginal misfit among all investigated groups ($p < 0.05$). Individual intergroup comparison using Tukey Kramer post hoc test revealed that copings fabricated from the SLM technique displayed lower marginal misfit than DLP-Cast ($p = 0.001$) and CAD-CAM (88.943 ± 20.880 μm) ($p = 0.001$). However SLM copings showed higher but comparable misfits to Cast-LWT (47.861 ± 19.693 μm) samples ($p > 0.05$). The microCT images of the assessed samples are presented in Figure 3.

Table 2. Comparison of marginal misfit of the CoCr copings.

Study Group	Mean (μm)	SD	ANOVA *p* Value	Cast/LWT	CAD-CAM	SLM	DLP-Cast
Cast/LWT	47.861	19.693	0.001 $	1.000			
CAD-CAM	88.943	20.880		0.031 *	1.000		
SLM	27.193	8.519		0.074	0.001 *	1.000	
DLP-Cast	147.746	30.306		0.013 *	0.001 *	0.001 *	1.000

* Statistical significant difference among groups shown in corresponding rows and columns using Tukey Kramer post hoc test; $ Statistically significant difference using ANOVA.

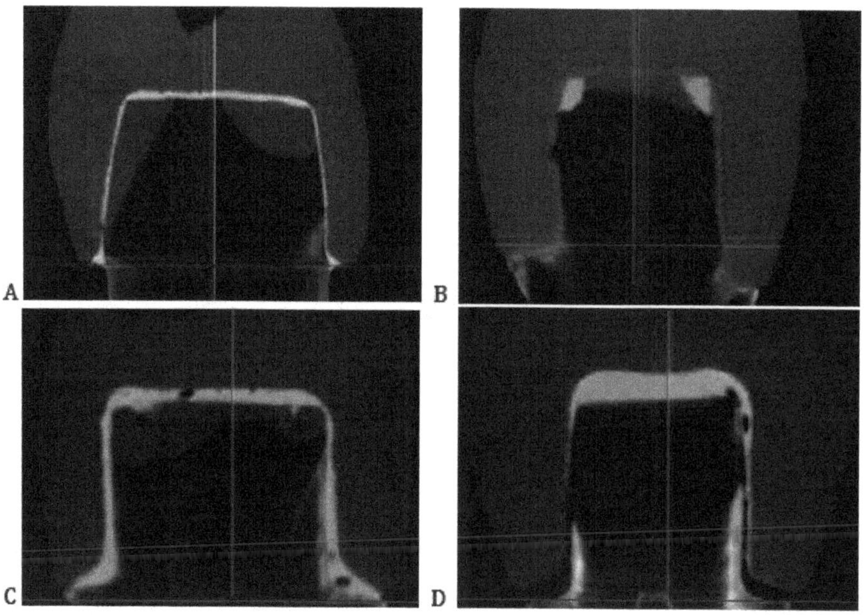

Figure 3. MicroCT images of the assessed samples in groups (**A**) Cast/LWT, (**B**) CAD-CAM, (**C**) DLP-Cast, and (**D**) SLM.

Figures 4–7 present the correlation between Ra and marginal misfit in CAD-CAM, Cast-LWT, SLM, and DLP-Cast study samples respectively. It was observed that Ra influences the marginal misfit in CAD-CAM (81.7%), SLM (94.8%), and DLP Cast (98.6%) technique-fabricated copings. ($p < 0.01$). Whereas, copings which are fabricated from CAST-LWT technique did not display any significant effect of Ra on the marginal misfit on the specimens of this group ($p = 0.435$), as displayed in the correlation plot (Figure 5).

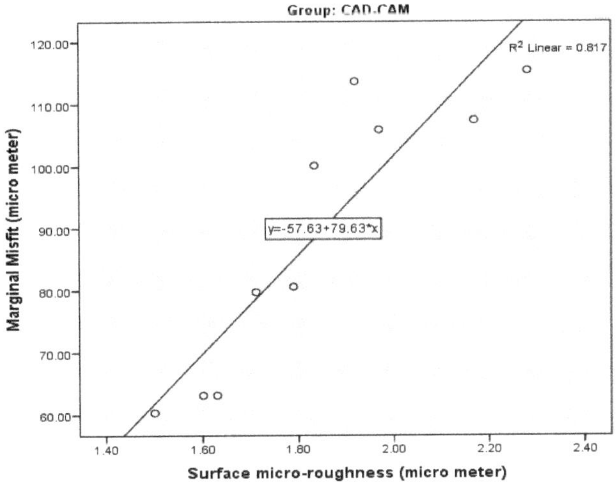

Figure 4. Showing a positive correlation between surface micro roughness and marginal misfit in CAD-CAM samples. R^2 showed 81.7% variation in marginal misfit explained by surface micro roughness; *p*-value was less than 0.01, and therefore statistically significant.

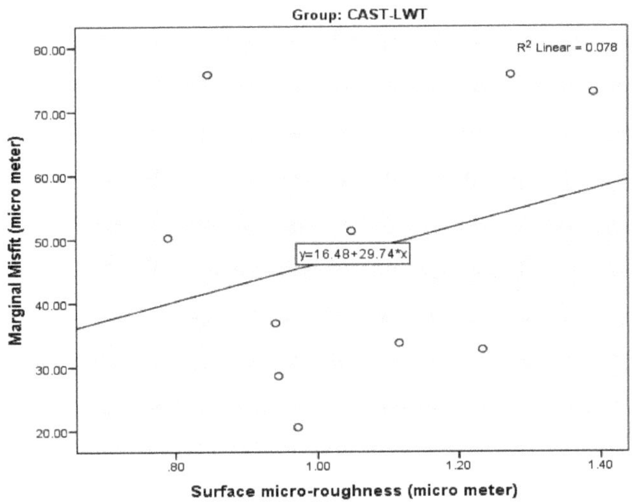

Figure 5. Showing a positive correlation between surface micro roughness and marginal misfit in CAST-LWT samples. R^2 showed 7.8% variation in marginal misfit explained by surface micro roughness; *p*-value 0.435, and therefore statistically insignificant.

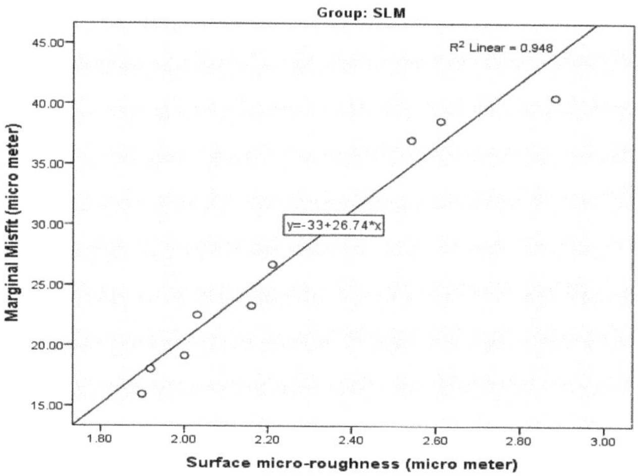

Figure 6. Showing a positive correlation between surface micro roughness and marginal misfit in SLM samples. R^2 showed 94.8% variation in marginal misfit explained by surface micro roughness; *p*-value was less than 0.01, and therefore statistically significant.

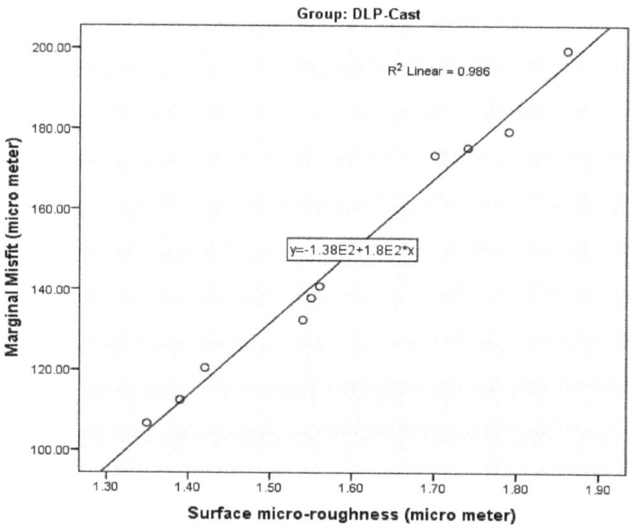

Figure 7. Showing a positive correlation between surface micro roughness and marginal misfit in DLP Cast samples. R^2 showed 98.6% variation in marginal misfit explained by surface micro roughness; *p*-value was less than 0.01, and therefore statistically significant.

4. Discussion

The present in vitro study was based on the hypothesis that there is no difference on Ra and marginal misfit of CoCr copings manufactured by conventional (Cast-LWT) and contemporary techniques (CAD-CAM, SLM, DLP-Cast). However, the existing study revealed that the additive technique of SLM showed lower misfit and high roughness. Whereas, the DLP-Cast specimen displayed higher internal misfit and lower Ra than the conventional technique. Therefore, the postulated hypothesis was rejected. These outcomes can be attributed to the number of steps, scanning and software limitations, and non-optimal parameters.

Contemporary CoCr alloys have gained popularity as compared with conventional gold alloys due to improved corrosion resistance and low cost [23]. The available literature showed multiple techniques, i.e., the direct-view measurement technique, the silicone replica technique, and cross-sectioning to measure the marginal misfit of fabricated copings [26]. The micro-computed tomography (CT), on the other hand, is comparatively an advanced method to assess the fit of indirect restoration through processing of scanned specimens slices, reconstructing the assembly by using software and gaging the misfit [27]. Similarly, in the present study the LPM was used to evaluate the Ra of CoCr copings. It is an optical system which is used to scan comparatively larger surface areas. It exhibits advantages over other techniques, i.e., determining surface characteristics such as the height of the largest profile projection and the depth of the largest profile depression [28].

In the present study it was found that SLM fabrication technique displayed lower marginal misfit than the other investigated groups. The better performance of the SLM technique is in line with the results of the study conducted by Fathi et al. [29]. Although, in the study by Fathi et al., assessments were performed using a silicone impression technique [29]. This can be explained by the fact that a lesser number of steps involved in the technique contributes to the precisely fitting copings as compared to other tested techniques. Moreover, copings fabricated from Cast-LWT displayed higher marginal misfit as compared with SLM copings. Multiple factors explain the increased misfit of Cast-LWT compared with SLM samples. Multiple steps are involved in the production of prosthesis through Cast-LWT and each step poses a risk of incorporating error thus compromising prosthetic fit [16]. Moreover, the accuracy of cast coping obtained through Cast-LWT depends on the accuracy of wax pattern and technical accuracy, i.e., wax composition, tank and block temperature, time specified for cooling of the wax pattern, and the firing temperature necessary to achieve desirable outcomes [30]. In addition, distortion of inlay casting wax, its shrinkage, and high investment expansion affects the precision of copings fabricated with Cast-LWT [31].

In the present study, DLP-Cast copings exhibited the highest marginal misfit compared with all other techniques. Marginal fit is highly dependent on the material properties utilized to fabricate copings in the 3D printer [32]. Moreover, resin used in the DLP-Cast technique undergoes polymerization shrinkage which generates stress, resulting in distortion of internal and marginal misfit [33]. In addition, the effect of scan spray when scanning the models cannot be overlooked [4]. These findings are in accordance with the study conducted by Kim at el., which used M-One printer, resulting in greater marginal misfit of DLP-Cast than other groups tested [25]. Kim et al. assessed the misfit using the weight of the silicone material and a digital microscope with sectioned specimens. Furthermore, it was also found that CAD-CAM displayed a significantly higher marginal misfit compared with Cast-LWT and SLM copings. This may be because correctness of internal and the marginal fit of copings produced through the subtractive technique depends on bur size. Any discrepancy in selection and size of burs relative to the size of coping results in compromised marginal properties [34]. Moreover, scanning and software system limitations related to finite resolution plausibly explicate the findings as these might result in margins with lower fit accuracy [35]. Many studies have compared the marginal or internal fit of coping and crowns fabricated from CAD-CAM and Cast-LWT [5,12,36]. Yet, it is challenging to conclude the findings of those studies due to variations in sample size, methods adopted to measure the marginal or internal gap, the type of cement used, and the CAD-CAM systems chosen. However, results of multiple previous studies are in line with the finding of the present study and displayed larger marginal or internal misfit for CAD-CAM fabricated prostheses than conventional techniques [12,37,38].

The influence of the different fabrication techniques on the Ra of CoCr copings have been addressed, proposing that Ra varies with the type of manufacturing technique [30]. Available literature also suggested that Ra values of any indirect prosthesis should be at least 0.2 μm [39]. Ra is suggested to influence retention of restorations, but its effect on the marginal fit is not clear. In the present study, it was found that all fabrication techniques demonstrated Ra higher than the recommended threshold. Moreover, the difference in Ra values among different AM techniques, i.e., DLP-Cast and SLM, can be explained by different parameters adopted during fabrication [40]. It is suggested that the use of non-optimal parameters in any of the AM techniques result in porosity and an increase in the Ra of the prosthesis [41]. In the authors' opinion, decreased marginal misfit in SLM-fabricated copings may be due to the highest surface roughness obtained in this group. Similarly, surface roughness of copings fabricated from the CAD-CAM subtractive technique depends on the cutting speed and depth of the cutting [42]. Moreover, LWT copings displayed the lowest Ra score; this may be due to the favorable inlay wax surface properties along with strict adherence to manufacturer guidelines aiding low Ra [43].

The study showed that additive manufacturing methods like SLM and DLP-cast showed varying success in producing marginally accurate and smooth metal copings. The findings suggest that SLM copings have a good marginal fit and high roughness; however, DLP-Cast specimens display low roughness but a poor marginal fit. Therefore, both of these techniques need further development and can only be applied in limited clinical contexts. The findings of the study should be interpreted in light of the possible study limitations. In the present study, copings were manufactured under ideal circumstances and controlled conditions. In addition, the outcomes of the study are limited to the materials and techniques employed in the experiments. Therefore, to translate the findings of the present study into clinical recommendations, randomized clinical trials assessing the fit and adaptation of CoCr copings fabricated with additive manufacturing techniques are warranted.

5. Conclusions

The SLM technique showed an improved coping marginal fit, but high roughness compared with controls. By contrast, the DLP-Cast specimens had smooth surfaces with poor marginal fits. Therefore, the application of additive manufacturing methods (SLM and DLP-Cast) in the fabrication of CoCr metal alloy coping needs further development.

Author Contributions: Conceptualization, A.S.A., F.V., T.A. and S.A.-S.; methodology, A.S.A., A.A., T.A. and M.E.; validation, A.S.A., S.A.-S., A.A., M.E. and M.K.A.; formal analysis, A.A., A.S.A., T.A., M.K.A. and M.E.; investigation, A.S.A., T.A., M.K.A., M.E., A.M.A. and F.V.; resources, T.A., S.A. and A.M.A.; data curation, M.K.A., S.A. and F.V.; writing—original draft preparation, F.V., S.A.-S. and S.A.; writing—review and editing, T.A., A.M.A. and A.A.; supervision, F.V., T.A. and A.S.A.; funding acquisition, T.A. All authors have read and agreed to the published version of the manuscript.

Funding: The authors are grateful to the Researchers supporting project at King Saud University for funding through Researchers supporting project No. (RSP-2019-44).

Institutional Review Board Statement: No ethical approval was required as patients or human or animal tissues were not involved in the study.

Informed Consent Statement: Not applicable.

Data Availability Statement: The data is available on contact from the corresponding author.

Acknowledgments: The authors are grateful to the Researchers supporting project at King Saud University for funding through Researchers supporting project No. (RSP-2019-44).

Conflicts of Interest: The authors declare no conflict of interest.

References

1. Vohra, F.; Bin Ismail, I.H.; Al-Rifaiy, M.Q.; Abduljabbar, T.; Abualsaud, H.; Ali, M.; Abu Hasan, M.I. Influence of veneering technique and veneer-coping thickness on fracture toughness of implant retained veneered Y-TZP zirconia crowns. *J. Adhes. Sci. Technol.* **2017**, *31*, 1758–1767. [CrossRef]
2. Samer, M.S.; Faraz, Q.; Al-Dubai, S.A.R.; Vohra, F.; Abdullah, H.; Taiyeb-Ali, T.B.; Saub, R. Clinical outcomes and predictors of satisfaction in patients with improved lithium disilicate all-ceramic crowns. *Med. Princ. Pract.* **2017**, *26*, 470–479. [CrossRef] [PubMed]

3. Al-Aali, K.A.; Bin Shuwaish, M.S.; Alhenaki, A.M.; Al Ahdal, K.; Al Deeb, L.; Maawadh, A.M., AlHelal, A.; Alshehri, H.; Vohra, F.; Abduljabbar, T. Influence of milling systems and marginal configurations on the fit of yttrium stabilized tetragonal zirconia polycrystals (Y-TZP)'copings. *J. Appl. Biomater. Funct. Mater.* **2020**, *18*, 2280800020924514.
4. Al-Aali, K.A.; Alhamdan, R.S.; Maawadh, A.M.; Vohra, F.; Abduljabbar, T. Influence of contemporary CAD-CAM milling systems on the fit and adaptation of partially stabilized Zirconia fixed partial dentures. *Pak. J. Med. Sci.* **2021**, *37*, 45–49. [PubMed]
5. Al Deeb, L.; Al Ahdal, K.; Alotaibi, G.; Alshehri, A.; Alotaibi, B.; Alabdulwahab, F.; Al Deeb, M.; AlFawaz, Y.F.; Vohra, F.; Abduljabbar, T. Marginal Integrity, Internal Adaptation and Compressive Strength of 3D Printed, Computer Aided Design and Computer Aided Manufacture and Conventional Interim Fixed Partial Dentures. *J. Biomater. Tissue Eng.* **2019**, *9*, 1745–1750. [CrossRef]
6. Bin-Shuwaish, M.; AlFawaz, Y.F.; AlGamaiah, H.A.; AlSani, A.S.; Abobakr, I.B.; Alzahrani, K.M.; Almutairi, B.; Attar, E.A.; Vohra, F.; Abduljabbar, T. Technical accuracy of dental laboratories in the quality and shade matching of porcelain fused to metal crowns: An in vitro study. *Int. J. Environ. Res. Public Health* **2021**, *18*, 2722. [CrossRef]
7. Hunt, L.B. The long history of lost wax casting—Over five thousand years of art and craftsmanship. *Gold Bull.* **1980**, *13*, 63–79. [CrossRef]
8. Vohra, F.; Shuwaish, M.B.; Al Deeb, M.; Alhamdan, R.; Alotaibi, N.; Abduljabbar, T. Comparison of failure loads and compressive stress in Press on metal and Press on Y-TZP copings. *Pak. J. Med. Sci.* **2020**, *36*, 1645–1650. [CrossRef] [PubMed]
9. McCoy, T. Lost wax casting technique for metal crown fabrication. *J. Vet. Dent.* **2014**, *31*, 126–132. [CrossRef] [PubMed]
10. Carrabba, M.; Vichi, A.; Vultaggio, G.; Pallari, S.; Paravina, R.; Ferrari, M. Effect of finishing and polishing on the surface roughness and gloss of feldspathic ceramic for chairside CAD/CAM systems. *Oper. Dent.* **2017**, *42*, 175–184. [CrossRef]
11. Rödiger, M.; Schneider, L.; Rinke, S. Influence of material selection on the marginal accuracy of CAD/CAM-fabricated metal-And all-ceramic single crown copings. *BioMed Res. Int.* **2018**, *2018*, 2143906. [CrossRef]
12. Ahrberg, D.; Lauer, H.C.; Ahrberg, M.; Weigl, P. Evaluation of fit and efficiency of CAD/CAM fabricated all-ceramic restorations based on direct and indirect digitalization: A double-blinded, randomized clinical trial. *Clin. Oral Investig.* **2016**, *20*, 291–300. [CrossRef]
13. Rajan, B.N.; Jayaraman, S.; Kandhasamy, B.; Rajakumaran, I. Evaluation of marginal fit and internal adaptation of zirconia copings fabricated by two CAD—CAM systems: An in vitro study. *J. Indian Prosthodont. Soc.* **2015**, *15*, 173–178. [CrossRef] [PubMed]
14. Mansour, F.K.; Ibrahim, R.M.; Mansour, H.; Hamdy, A.M. Assessment of internal fit and micro leakage of conventionally fabricated ceramometallic restoration versus CAD wax and press veneering (in-vitro study). *BDJ Open* **2021**, *7*, 17. [CrossRef]
15. Farjood, E.; Vojdani, M.; Torabi, K.; Khaledi, A.A.R. Marginal and internal fit of metal copings fabricated with rapid prototyping and conventional waxing. *J. Prosthet. Dent.* **2017**, *117*, 164–170. [CrossRef] [PubMed]
16. Lövgren, N.; Roxner, R.; Klemendz, S.; Larsson, C. Effect of production method on surface roughness, marginal and internal fit, and retention of cobalt-chromium single crowns. *J. Prosthet. Dent.* **2017**, *118*, 95–101. [CrossRef] [PubMed]
17. Zocca, A.; Colombo, P.; Gomes, C.M.; Günster, J. Additive Manufacturing of Ceramics: Issues, Potentialities, and Opportunities. *J. Am. Ceram. Soc.* **2015**, *98*, 1983–2001. [CrossRef]
18. Gholamrezaei, K.; Vafaee, F.; Afkari, B.F.; Firouz, F.; Seif, M. Fit of cobalt-chromium copings fabricated by the selective laser melting technology and casting method: A comparative evaluation using a profilometer. *Dent. Res. J.* **2020**, *17*, 200–207.
19. Ren, L.; Memarzadeh, K.; Zhang, S.; Sun, Z.; Yang, C.; Ren, G.; Allaker, R.P.; Yang, K. A novel coping metal material CoCrCu alloy fabricated by selective laser melting with antimicrobial and antibiofilm properties. *Mater. Sci. Eng. C* **2016**, *67*, 461–467. [CrossRef]
20. Li, X.; Zhong, H.; Zhang, J.; Duan, Y.; Li, J.; Jiang, D. Fabrication of zirconia all-ceramic crown via DLP-based stereolithography. *Int. J. Appl. Ceram. Technol.* **2019**, *17*, 844–853. [CrossRef]
21. Chen, Z.; Li, Z.; Li, J.; Liu, C.; Lao, C.; Fu, Y.; Liu, C.; Li, Y.; Wang, P.; He, Y. 3D printing of ceramics: A review. *J. Eur. Ceram. Soc.* **2019**, *39*, 661–687. [CrossRef]
22. Pettenò, D.; Schierano, G.; Bassi, F.; Bresciano, M.E.; Carossa, S. Comparison of Marginal Fit of 3 Different Metal-Ceramic Systems: An In Vitro Study. *Int. J. Prosthodont.* **2001**, *13*, 405–408.
23. Kim, M.-J.; Choi, Y.-J.; Kim, S.-K.; Heo, S.-J.; Koak, J.-Y.; Weber, F.E. Marginal Accuracy and Internal Fit of 3-D Printing Laser-Sintered Co-Cr Alloy Copings. *Materials* **2017**, *10*, 93. [CrossRef] [PubMed]
24. Zeltser, C.; Lewinstein, I.; Grajower, R. Fit of crown wax patterns after removal from the die. *J. Prosthet. Dent.* **1985**, *53*, 344–346. [CrossRef]
25. Venkatesh, K.V.; Nandini, V.V. Direct metal laser sintering: A digitised metal casting technology. *J. Indian Prosthodont. Soc.* **2013**, *13*, 389–392. [CrossRef]
26. Ram, S.; Ranadive, N.; Nadgere, J. Microcomputed tomography a noninvasive method to evaluate the fit of a restoration as compared to conventional replica technique. *J. Indian Prosthodont. Soc.* **2019**, *19*, 233–239. [CrossRef]
27. Pelekanos, S. Micro-CT evaluation of the marginal fit of different In-Ceram alumina copings. *Eur. J. Esthet. Dent. Off. J. Eur. Acad. Esthet. Dent.* **2009**, *4*, 278–292.
28. Kulik, E.A.; Calahan, P. Laser Profilometry of Polymeric Materials. *Cells Mater.* **1997**, *7*, 3.
29. Fathi, H.M.; Al-Masoody, A.; Johnson, A. The accuracy of fit of crowns made from wax patterns produced conventionally (hand formed) and via CAD/CAM technology. *Eur. J. Prosthodont. Restor. Dent.* **2016**, *24*, 10–17. [CrossRef] [PubMed]
30. Sravani, K. To Evaluate And Compare The marginal fit and internal adaptation of cast copings fabricated with pattern wax and autopolymerized pattern resin materials. *IOSR J. Dent. Med Sci. (IOSR-JDMS)* **2020**, *19*, 54–66.

31. Oguz, E.I.; Kılıçarslan, M.A.; Özgür, M.E.; Orhan, K.; Shujaat, S. Comparison of Marginal Adaptation of Different Resin-Ceramic CAD/CAM Crowns: An In Vitro Study. *J. Adv. Oral Res.* **2021**, *12*, 112–118. [CrossRef]
32. Cui, K.; Shang, X.; Luo, C.; Shen, Z.; Gao, H.; Xiong, G. A kind of accuracy improving method based on error analysis and feedback for DLP 3D Printing. In Proceedings of the IEEE international conference on service operations and logistics, and Informatics 2019 (SOLI 2019), Zhengzhou, China, 6–8 November 2019; pp. 5–9.
33. Kim, S.Y.; Shin, Y.S.; Jung, H.D.; Hwang, C.J.; Baik, H.S.; Cha, J.Y. Precision and trueness of dental models manufactured with different 3-dimensional printing techniques. *Am. J. Orthod. Dentofac. Orthop.* **2018**, *153*, 144–153. [CrossRef]
34. Guncu, M.B.; Tuba Akcin, E.; Aktas, G.; Aslan, Y. Accuracy of CAD CAM manufacturing techniques on marginal and internal fit of 5-unit implant-supported Co-Cr frameworks. *Clin. Oral Implants Res.* **2018**, *29*, 162. [CrossRef]
35. Davidowitz, G.; Kotick, P.G. The Use of CAD/CAM in Dentistry. *Dent. Clin. N. Am.* **2011**, *55*, 559–570. [CrossRef]
36. Tan, P.L.; Gratton, D.G.; Diaz-Arnold, A.M.; Holmes, D.C. An in vitro comparison of vertical marginal gaps of CAD/CAM titanium and conventional cast restorations. *J. Prosthodont.* **2008**, *17*, 378–383. [CrossRef]
37. Pompa, G.; Di Carlo, S.; De Angelis, F.; Cristalli, M.P.; Annibali, S. Comparison of conventional methods and laser-assisted rapid prototyping for manufacturing fixed dental prostheses: An In vitro study. *BioMed Res. Int.* **2015**, *2015*, 318097. [CrossRef]
38. Alves de Carvalho, I.; Marques, T.; Araújo, F.; Azevedo, L.; Donato, H.; Correia, A. Clinical Performance of CAD/CAM Tooth-Supported Ceramic Restorations: A Systematic Review. *Int. J. Periodontics Restor. Dent.* **2018**, *38*, e68–e78. [CrossRef]
39. Bollen, C.M.; Lambrechts, P.; Quirynen, M. Comparison of surface roughness of oral hard materials to the threshold surface roughness for bacterial plaque retention: A review of the literature. *Dent. Mater.* **1997**, *13*, 258–269. [CrossRef]
40. Pleass, C.; Jothi, S. Influence of powder characteristics and additive manufacturing process parameters on the microstructure and mechanical behaviour of Inconel 625 fabricated by Selective Laser Melting. *Addit. Manuf.* **2018**, *24*, 419–431. [CrossRef]
41. Abate, K.M.; Nazir, A.; Chen, J.E.; Jeng, J.Y. Design, optimization, and evaluation of additively manufactured vintiles cellular structure for acetabular cup implant. *Processes* **2020**, *8*, 25. [CrossRef]
42. Koizumi, H.; Saiki, O.; Nogawa, H.; Hiraba, H.; Okazaki, T.; Matsumura, H. Surface roughness and gloss of current CAD/CAM resin composites before and after toothbrush abrasion. *Dent. Mater. J.* **2015**, *34*, 881–887. [CrossRef] [PubMed]
43. Vaidya, S.; Parkash, H.; Bhargava, A.; Gupta, S. Evaluation of the marginal fit of metal copings fabricated on three different marginal designs using conventional and accelerated casting techniques: An in vitro study. *Indian J. Dent. Res.* **2014**, *25*, 641–647. [CrossRef] [PubMed]

applied sciences

MDPI

Article

Influence of Root Canal Sealers and Obturation Techniques on Vertical Root Fracture Resistance. An In Vitro Experiment

Mazen F. Alkahtany [1], Khalid H. Almadi [1], Fahad A. Alahmad [2], Abdullah M. Alshehri [2], Abdulrahman A. AlSwayyed [2], Omar M. AlZahran [2], Ali AlHadan [3], Abdulaziz S. Almustafa [3], Fahim Vohra [4] and Tariq Abduljabbar [4,*]

1 Department of Restorative Dental Science, Division of Endodontics, College of Dentistry, King Saud University, Riyadh 11545, Saudi Arabia; malkahtany@ksu.edu.sa (M.F.A.); kalmadi@ksu.edu.sa (K.H.A.)

2 Dental Intern, College of Dentistry, King Saud University, Riyadh 11545, Saudi Arabia; fahad.alahmed96@gmail.com (F.A.A.); A.alshehri.dent@gmail.com (A.M.A.); Abo.mohannad1996@gmail.com (A.A.A.); Omardentt@gmail.com (O.M.A.)

3 Department of General Dentistry, College of Dentistry, King Saud University, Riyadh 11545, Saudi Arabia; aahdn87@gmail.com (A.A.); abdaziza@gmail.com (A.S.A.)

4 Department of Prosthetic Dental Science, College of Dentistry, King Saud University, Riyadh 11545, Saudi Arabia; fvohra@ksu.edu.sa

* Correspondence: tajabbar@ksu.edu.sa; Tel.: +966-1467-7444; Fax: +966-1467-8639

Abstract: The aim of the present study was to determine the vertical root fracture (VRF) resistance of roots obturated with TotalFill BC Sealer and AH Plus sealer using lateral condensation and single cone techniques in comparison to untreated controls. Sixty single rooted mandibular premolars were sectioned and divided into six groups. Ten teeth were left untreated (positive control-Gp 1) and fifty teeth were cleaned and shaped. Ten root specimens were left unfilled (negative control-Gp 2) and the remaining roots were divided into 4 groups. Gp 3, GP and AH Plus sealer (AH Plus) using the cold lateral compaction (LC) technique; Gp 4, GP and AH Plus using the Single Cone (SC) technique; Gp 5: TotalFill GP and TotalFill BC sealer using the LC technique; Gp 6: TotalFill GP and TotalFill BC sealer with SC. VRF was performed for all specimens using a universal testing machine. Analysis of variance (ANOVA) and Tukeys post-hoc multiple comparison test was used to compare the means among tested study groups. Group 1 (positive control) displayed the highest fracture resistance (946.61 ± 166.465 N); however, the lowest fracture strength was demonstrated by the specimens in group 2 (negative control) (433.31 ± 129.350 N). Specimens treated with AH plus using different obturation techniques (group 3 and 4) showed comparable outcomes ($p > 0.05$). Similarly, specimens treated with TotalFill BC sealer with different obturation techniques showed statistically similar outcomes ($p > 0.05$). It was also observed that specimens in groups 3, 4, 5 and 6 demonstrated comparable outcomes of fracture strength ($p > 0.05$). The use of TotalFill-BC sealer showed similar vertical root fracture resistance as AH plus sealer in root canal treated teeth. Use of total fill-BC and AH Plus sealer in root canal treatment showed vertical root fracture resistance comparable to untreated natural teeth (positive controls).

Keywords: sealers; vertical fracture; root filling; fracture resistance; obturation technique

Citation: Alkahtany, M.F.; Almadi, K.H.; Alahmad, F.A.; Alshehri, A.M.; AlSwayyed, A.A.; AlZahran, O.M.; AlHadan, A.; Almustafa, A.S.; Vohra, F.; Abduljabbar, T. Influence of Root Canal Sealers and Obturation Techniques on Vertical Root Fracture Resistance. An In Vitro Experiment. *Appl. Sci.* **2021**, *11*, 8022. https://doi.org/10.3390/app11178022

Academic Editor: Bruno Chrcanovic

Received: 27 July 2021
Accepted: 24 August 2021
Published: 30 August 2021

1. Introduction

An endodontically treated tooth is more prone to vertical root fracture (VRF) due to multiple contributing factors, i.e., caries or trauma, access cavity preparation, root canal instrumentation, lateral condensation force during obturation, preparation of post space and functional occlusal loading, which leads to failure [1]. VRF is considered one of the most common complications of endodontic treatment, eventually leading to extraction of the tooth [2]. It can occur during or after the root canal treatment (RCT) due to compromised tooth structure. Therefore, reinforcement of the remaining tooth structure is considered as an important aspect of endodontic treatment [3]. Available literature revealed that bonding of root filling with the radicular dentin strengthens the tooth as well as increases its fracture resistance [4,5].

Cleaned and shaped root canals obturated with Gutta-percha (GP) and root canal sealer is considered as the gold standard [6]. GP lack the capability to reinforce the weakened root structure because of their low modulus of elasticity [7]. Therefore, a root canal sealer is considered as a joint between a canal and a filling material. An ideal root canal sealer should possess the property of filling the apical and lateral empty spaces and irregularities between GP points and root dentin walls [8]. Moreover, it was also assumed that sealer, which exhibits adhesion to the canal dentin, results in reinforcement of the tooth and increases the fracture toughness against the occlusal load by conserving the integrity at the sealer–dentin interface [9]. Therefore, multiple research methodologies have introduced different canal sealers [10].

Among different sealers used recently, AH Plus (resin-based root canal sealer) is considered as a material of choice for canal obturation due to ease of handling, good mechanical properties, wettability and excellent sealing property [11]. Furthermore, it also exhibits less polymerization shrinkage, low solubility, and a high degree of stability on storage [12]. Studies have suggested that AH Plus sealer results in increased fracture resistance of endodontically treated teeth [12,13]. However, a few studies presented contradictory findings [9,14].

TotalFill BC Sealer on the other hand is a calcium-silicate based bio-ceramic root canal sealer. It is dispensed in a premixed ready-to-use injectable form that sets in the presence of water, possesses excellent antimicrobial ability, higher pH, and exceptional biocompatibility. TotalFill BC sealer can easily penetrate the dentinal tubules due to presence of nanoparticles. In addition, it does not shrink while setting and demonstrates excellent physical properties. Available literature revealed that calcium silicate based sealer reinforces the tooth structure against the VRF and strengthens the tooth. A study conducted by Ghoneim et al. [15] and Sağsen et al. [16] showed that bioceramic-based sealer (BC sealer) has the potential to increase the root fracture resistance [17]. In contrast, Celikten et al. reported that BC sealer has a significantly lower mean value for fracture than the control group [18].

From the available indexed literature it was found that sufficient data related to the effect of AH Plus sealer on the fracture resistance of root canal treated tooth are available. However, data on the influence of TotalFill BC sealer on fracture resistance of root treated teeth in comparison to commonly used sealers are limited. Therefore, it was hypothesized that there will be no difference in fracture resistance of roots obturated with TotalFill BC Sealer and AH Plus sealer. It was also hypothesized that there will be no difference in root fracture resistance when two different obturation techniques, i.e., lateral condensation and single cone technique, are employed. Thus, the aim of the present study was to determine the fracture resistance of roots obturated with TotalFill BC Sealer and AH Plus sealer using lateral condensation and single cone techniques in comparison to untreated controls.

2. Materials and Methods

2.1. Specimen Preparation

A total of sixty freshly extracted single rooted mandibular premolars were collected over a period of 3 months. All the teeth were stored in Hank's balanced salt solution (HBSS) till further use. All the specimens were cleaned, sterilized and made ready to examine under stereo-microscope (Stemi 2000-C, Zeiss, Wetzlar, Germany) at 50× magnification in order to exclude teeth with open apices, root caries, cracks and/or fractured roots. Any tooth with a resorption defect or previous root canal treatment was excluded. A pre-operative radiograph was taken both buccolingually and mesiodistally to confirm the presence of a single canal. De-coronation was performed at the level of CEJ with the help of a high speed hand-piece with a wheel diamond bur in order to standardize 13 mm of root length. The methodology outline is presented in Figure 1.

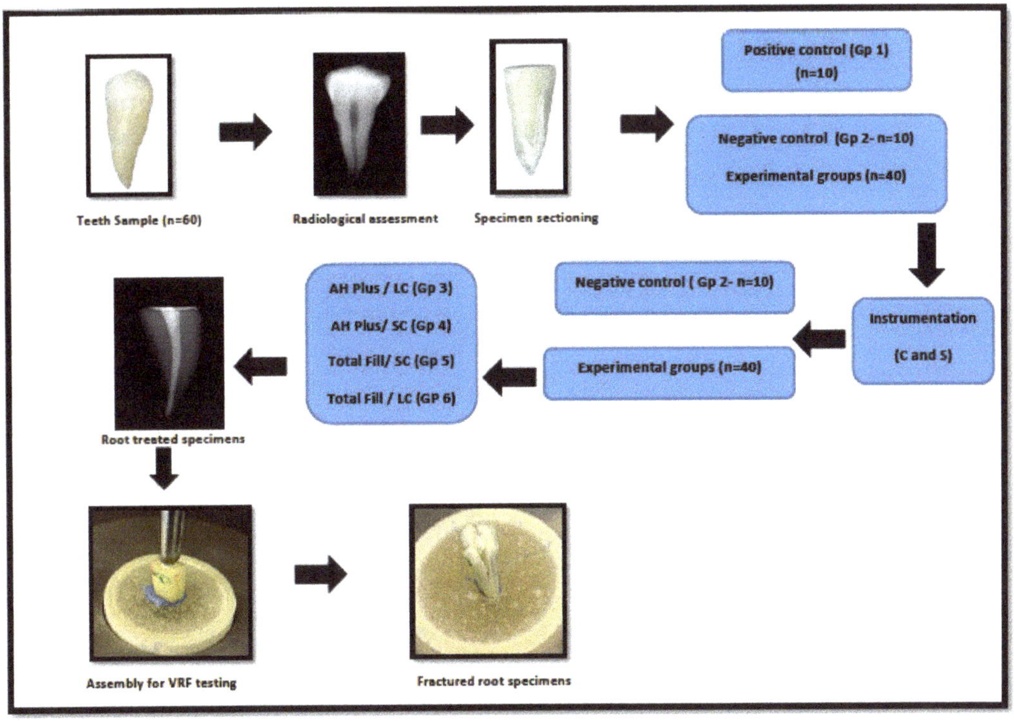

Figure 1. Flow chart for study methodology.

2.2. Treatment and Study Groups

Ten samples were kept un-instrumented as positive controls (group 1). A size #15 K patency file was used to determine the correct working length (WL). For fifty specimens, the k file was placed in the canal until it extrudeed the apex and then the file was pulled 1 mm shorter to the apex. Fifty specimens were then prepared using the crown-down technique by Profile rotary system up to #40/0.04 taper file (excluding positive control specimens-group 1). Between each instrumentation, constant irrigation was performed with 2.5% NaOCl (1 mL). After completion of the canal preparation, a final rinse was performed with 2.5% NaOCl (2 mL) for 1 min followed by 2 mL 17% EDTA for 1 min and 10 mL distilled water. Root canal surfaces were dried using paper points and were further divided into 5 groups (n = 10) based on instrumentation and obturation techniques.

Group 1 (Positive control): Samples were left un-instrumented and unfilled.

Group 2 (Negative control): Roots were instrumented and left unfilled.

Group 3: GP and AH Plus sealer using cold lateral compaction technique.

Group 4: GP and AH Plus sealer using Single Cone technique

Group 5: TotalFill GP and TotalFill BC sealer using cold lateral compaction technique

Group 6: TotalFill GP and TotalFill BC sealer using Single Cone technique.

After completing obturation, periapical radiographs were exposed to assess the quality of root filling. Any root with inadequate obturation was excluded and replaced with a newly prepared specimen. The access cavity was closed using temporary filling (3M™ Cavit™). The specimens were kept at 37 °C and 100% humidity for 2 weeks to allow complete setting of the sealer.

2.3. Specimen Testing

To simulate the periodontal attachment, the specimen's root surface was coated with a thin-layer of polyvinylsiloxane (PVS) impression material up to 2 mm apical to the coronal end of the root. All of the roots were mounted perpendicularly in a poly-vinyl ring filled with self-cure acrylic resin (Opti-Cryl, South Carolina, Columbia) exposing 2 mm from the root (cervical). Root fracture resistance was assessed using a universal testing machine (UTM) (Lloyds, LF, plus, Ametek Inc., Great Britain, UK). All of the samples were placed in the lower plate of the testing machine and in the upper part a custom-made metal spreader with a diameter of 0.8 mm was secured. The tip was oriented in the center of the canal orifice and force was applied vertically to the long axis of the root at a crosshead speed of 0.5 mm/min until root fracture. A drop in >25% of applied force was observed when the fractures occurred. The amount of load necessary for root fracture was recorded in Newtons (N).

2.4. Statistical Analysis

Data related to fracture strength was analyzed using Statistical package for the social sciences (SPSS V.25, IBM, New York, NY, USA). Normality was assessed using the Kolmogorov–Smirnov test. One-way analysis of variance (ANOVA) and Tukey's post-hoc multiple comparison test were used to compare the means among tested study groups.

3. Results

The mean and standard deviations of fracture resistance among the investigated groups are presented in Table 1.

Table 1. Means and SD of vertical root fracture resistance among study groups.

Study Group	Mean	SD	ANOVA (p Value)
1. Positive control	946.61	166.46	
2. Negative control	433.31	129.35	
3. GP-AH Plus-LC	733.71	232.57	$p < 0.05$
4. GP-AH Plus-SC	752.77	120.58	
5. Totalfill GP-BC-LC	701.11	65.29	
6. Totalfill GP-BC-SC	797.46	204.55	

SD. Standard deviation, LC. Cold lateral compaction, SC. Single cone, GP. Gutta Percha.

The Kolmogorov–Smirnov test displayed normal distribution of data. Specimens of group 1 (positive control) displayed the highest fracture resistance (946.61 ± 166.465 N); however, the lowest fracture strength was demonstrated by the specimens in group 2 (negative control) (433.31 ± 129.350 N). ANOVA revealed that fracture strengths among study groups were statistically significant ($p \leq 0.05$).

Individual comparisons among all the investigated groups established that group 2 displayed significantly lower fracture strength among all experimental groups. Group 1 (positive control) showed higher fracture strength than groups 2 ($p < 0.01$) and 5 ($p = 0.0$); however, it was comparable to groups 3 ($p > 0.05$), 4 ($p = 0.10$) and 6 ($p = 0.33$), respectively. Specimens treated with AH plus using different obturation techniques (group 3 and 4) showed comparable outcomes ($p = 0.99$). Similarly, specimens treated with TotalFill BC sealer with different obturation techniques showed statistically similar outcomes ($p = 0.77$). It was also observed that specimens in groups 3, 4, 5 and 6 demonstrated comparable outcomes of fracture strength ($p > 0.05$) (Table 2).

Table 2. Statistical comparison of root fracture resistance among study groups (Tukey's post hoc test).

Study Groups	Group Comparison	p Value
1. Positive control	1 vs 2	0.0000
	1 vs 3	0.0541
	1 vs 4	0.1003
	1 vs 5	0.0166
	1 vs 6	0.3300
2. Negative control	2 vs 3	0.0017
	2 vs 4	0.0007
	2 vs 5	0.0069
	2 vs 6	0.0001
3. GP-AH Plus-LC	3 vs 4	0.9998
	3 vs 5	0.9976
	3 vs 6	0.9508
4. GP-AH Plus-SC	4 vs 5	0.9800
	4 vs 6	0.9896
5. Totalfill GP-BC-LC 6. Totalfill GP-BC-SC	5 vs 6	0.7713

LC. Cold lateral compaction, SC. Single cone, GP. Gutta Percha.

4. Discussion

The present study was based on the hypothesis that there will be no difference in fracture resistance of root canals obturated with TotalFill BC Sealer and AH Plus sealer. It was also hypothesized that there will be no difference in fracture resistance when two different obturation techniques, i.e., lateral condensation and the single cone technique were used. Thus, the postulated hypothesis was accepted as experimental groups in which root canals were obturated using different sealers (TotalFill sealer and AH Plus sealer) and different techniques (lateral condensation technique and Single Cone technique) showed comparable fracture resistance outcomes. A multitude of reasons are responsible for such outcomes, including the adhesive properties of the sealer, bioactivity of sealers, instrument type and instrumentation methods, and root dentin anatomy.

In the present study, in order to avoid inter-operator variability, a single operator performed all procedures including root canal preparation, irrigation and obturation. Similarly, for root canal preparation, the crown down technique was adopted as it allows debris to be expelled from the canal orifice [19]. Moreover, further standardization was achieved by using AH Plus sealer, which is considered as a gold standard root canal sealer in dentistry [20].

A sealer is conceived as a joint created between radicular dentin and the root filling material [13]. Adhesion between an endodontic sealer and root dentin serves two important purposes [21]. Primarily, the root canal sealer gives a superior seal, which prevents coronal and apical leakage [22]. Secondly, it inhibits filling material displacement during restorative procedures. In the present study, it was found that the mean value of fracture loads of the positive control (433.31 ± 129.350 N) was significantly lower than all other groups tested. The above findings can be associated with loss of radicular dentin thickness (RDT) and moisture due to canal instrumentation and the reinforcement effect by both sealers [23,24].

It was also established that positive control specimens, in which no canal instrumentation performed displayed a fracture strength comparable to group 3 (AH plus sealer + Lateral condensation technique) (733.71 ± 232.572 N), group 4 (AH plus sealer + Single Cone technique) (752.77 ± 120.587 N), and group 6 (TotalFill sealer + Single Cone technique) (797.46 ± 204.557 N). There are various justifications that are accredited to such an outcome. AH Plus sealer, being epoxy resin-based, unveils some desired proprieties, i.e., adhesion by forming a covalent bond between the open epoxide ring and exposed amino acids in the collagen [25]. Moreover, AH Plus possesses an excellent penetration ability into the surface micro-irregularities due its creeping property, which results in increased fracture strength [11,26]. This finding is in line with the results of the earlier stated studies by Sağsen et al. and Topçuoğlu et al., which suggested that obturation with AH plus root canal sealers are able to resist the fracture load equivalent to the sound tooth structure in which no canal preparation and filling was performed [3,16]. Similarly, the comparable fracture resistance demonstrated by the BC sealer in group 6 specimens to the positive control group can be explained by its property to produce hydroxyapatite, which leads to increased chemical bonding of sealer to the canal dentinal walls [11,27]. In addition, the presence of small "nanoparticles" and their ability to penetrate deeply into isthmuses, accessory canals and canal irregularities also justifies the higher fracture strength of Total-Fill sealer [5,28]. This finding is in accordance with the outcomes of several studies that proposed that BC-based sealers were able to increase the fracture resistance comparable to that of the intact tooth [29,30].

On the other hand, it was also found that root filling with the lateral condensation technique and Total fill root canal sealer is able to increase the fracture strength of the specimens but is not comparable to the un-instrumented sound tooth. Similar results were noted in the study conducted by Saw and Messer [31]. This finding can be explained by the fact that the finger spreader used for the lateral condensation technique generates stress on the canal wall that may weaken the tooth resulting in less fracture resistance [32]. Spreader design and applied forces are suggested as the contributing factors to the appearance of vertical root fractures during lateral compaction [31]. Moreover, the comparable fracture resistance among all the groups, which were sealed using TotalFill sealer and AH Plus sealer, further suggested that the difference in fracture load may be due to the difference in methodologies used [33]. Sağsen et al. and Mohammed & Al-Zaka. in their studies revealed that difference in fracture strength among different sealer groups was due to variation in the technique opted for obturation [11,16].

The present in vitro study presented some inherent limitations. The diameter of the root was not standardized in the present study, which has a potential influence on the fracture resistance of the tooth. Moreover, the impact of canal shape cannot be overlooked as more tapered canals results in more dentin removal from the canal resulting in weakening of the specimens. Furthermore, the amount of dentinal tubules present in each specimen also influences the outcomes of root fracture resistance. As the present study was an in vitro experiment, more clinical-based studies should be conducted to validate the findings of the present study for clinical applications.

5. Conclusions

The use of TotalFill-BC sealer showed similar vertical root fracture resistance reinforcing effect as AH plus sealer in root canal treated teeth. Use of total fill-BC and AH Plus sealer in root canal treatment showed vertical root fracture resistance comparable to untreated natural teeth. Use of different obturation techniques (Single cone and lateral condensation technique) in the presence of sealers (AH plus and TotalFill) did not show a significant influence on vertical root fracture resistance.

Author Contributions: Conceptualization, M.F.A., K.H.A., F.A.A., A.M.A. and A.S.A.; methodology, M.F.A., F.A.A., K.H.A. and F.V.; validation, F.V., A.M.A. and T.A.; formal analysis, A.S.A., A.M.A., A.A.A. and K.H.A.; investigation, M.F.A., A.A.A. and F.V.; resources, T.A. and A.A.A.; data curation, F.V., K.H.A., O.M.A., A.A.A. and F.A.A.; writing—original draft preparation, F.V.; writing—review and editing, T.A. and F.V.; supervision, F.V., A.S.A., O.M.A. and A.M.A.; funding acquisition, T.A.; surgical assistance and data collection, revisions, A.A. All authors have read and agreed to the published version of the manuscript.

Funding: The authors are grateful to the Researchers supporting project at King Saud University for funding through Researchers supporting project No. (RSP-2019-44).

Institutional Review Board Statement: This study was submitted, reviewed, and approved by, Specialist Dental practice and Research Centre, Riyadh, Saudi Arabia (UDCRC/009-20). The ethical standards of the 1964 Helsinki declaration and national and/or institutional research committee were strictly followed while performing all the procedures. Additional information on the study was provided verbally by the study investigator or in a written format.

Informed Consent Statement: Consent was taken from individuals at teeth extraction.

Data Availability Statement: The data is available on contact from the corresponding author.

Acknowledgments: The authors are grateful to the Researchers supporting project at King Saud University for funding through Researchers supporting project No. (RSP-2019-44).

Conflicts of Interest: The authors declare no conflict of interest.

References

1. Touré, B.; Faye, B.; Kane, A.W.; Lo, C.M.; Niang, B.; Boucher, Y. Analysis of reasons for extraction of endodontically treated teeth: A prospective study. *J. Endod.* **2011**, *37*, 1512–1515. [CrossRef]
2. Bhat, S.; Hegde, S.; Rao, A.; Shaji Mohammed, A. Evaluation of resistance of teeth subjected to fracture after endodontic treatment using different root canal sealers: An in vitro study. *J. Indian Soc. Pedod. Prev. Dent.* **2012**, *30*, 305–309. [CrossRef] [PubMed]
3. Topçuoğlu, H.S.; Tuncay, Ö.; Karataş, E.; Arslan, H.; Yeter, K. In vitro fracture resistance of roots obturated with epoxy resin-based, mineral trioxide aggregate-based, and bioceramic root canal sealers. *J. Endod.* **2013**, *39*, 1630–1633. [CrossRef] [PubMed]
4. Uzunoglu-Özyürek, E.; Küçükkaya Eren, S.; Karahan, S. Effect of root canal sealers on the fracture resistance of endodontically treated teeth: A systematic review of in vitro studies. *Clin. Oral Investig.* **2018**, *22*, 2475–2485. [CrossRef]
5. Ribeiro, F.C.; Souza-Gabriel, A.E.; Marchesan, M.A.; Alfredo, E.; Silva-Sousa, Y.T.C.; Sousa-Neto, M.D. Influence of different endodontic filling materials on root fracture susceptibility. *J. Dent.* **2008**, *36*, 69–73. [CrossRef]
6. Karapinar Kazandag, M.; Sunay, H.; Tanalp, J.; Bayirli, G. Fracture resistance of roots using different canal filling systems. *Int. Endod. J.* **2009**, *42*, 705–710. [CrossRef]
7. Punjabi, M.; Dewan, R.G.; Kochhar, R. Comparative evaluation of fracture resistance of root canals obturated with four different obturating systems. *J. Conserv. Dent.* **2017**, *20*, 445–450. [CrossRef] [PubMed]
8. Zamin, C.; Silva-Sousa, Y.T.C.; Souza-Gabriel, A.E.; Messias, D.F.; Sousa-Neto, M.D. Fracture susceptibility of endodontically treated teeth. *Dent. Traumatol.* **2012**, *28*, 282–286. [CrossRef]
9. Sandikçi, T.; Kaptan, R.F. Comparative evaluation of the fracture resistances of endodontically treated teeth filled using five different root canal filling systems. *Niger. J. Clin. Pract.* **2014**, *17*, 667–672. [CrossRef]
10. Küçükkaya Eren, S.; Uzunoğlu-Özyürek, E.; Karahan, S. Influence of reciprocating and rotary instrumentation on microbial reduction: A systematic review and meta-analysis of in vitro studies. *Restor. Dent. Endod.* **2021**, *46*, e19. [CrossRef]
11. Mohammed, Y.T.; Al-Zaka, I.M. Fracture resistance of endodontically treated teeth obturated with different root canal sealers (A comparative study). *J. Contemp. Dent. Pract.* **2020**, *21*, 490–493. [CrossRef] [PubMed]
12. Mandava, J.; Chang, P.C.; Roopesh, B.; Faruddin, M.G.; Anupreeta, A.; Uma, C. Comparative evaluation of fracture resistance of root dentin to resin sealers and a MTA sealer: An in vitro study. *J. Conserv. Dent.* **2014**, *17*, 53. [CrossRef]
13. Phukan, A.H.; Mathur, S.; Sandhu, M.; Sachdev, V. The effect of different root canal sealers on the fracture resistance of endodontically treated teeth-in vitro study. *Dent. Res. J.* **2017**, *14*, 382. [CrossRef] [PubMed]
14. Dibaji, F.; Afkhami, F.; Bidkhori, B.; Kharazifard, M.J. Fracture Resistance of Roots after Application of Different Sealers. *Iran. Endod. J.* **2017**, *12*, 50. [CrossRef] [PubMed]
15. Ghoneim, A.G.; Lutfy, R.A.; Sabet, N.E.; Fayyad, D.M. Resistance to fracture of roots obturated with novel canal-filling systems. *J. Endod.* **2011**, *37*, 1590–1592. [CrossRef]
16. Sagsen, B.; Er, O.; Kahraman, Y.; Akdogan, G. Resistance to fracture of roots filled with three different techniques. *Int. Endod. J.* **2007**, *40*, 31–35. [CrossRef]
17. Yaman, S.D.; Alaçam, T.; Yaman, Y. Analysis of stress distribution in a vertically condensed maxillary central incisor root canal. *J. Endod.* **1995**, *21*, 321–325. [CrossRef]
18. Celikten, B.; Uzuntas, C.F.; Gulsahi, K. Resistance to fracture of dental roots obturated with different materials. *Biomed. Res. Int.* **2015**, *2015*. [CrossRef]
19. Jindal, R.; Singh, S.; Gupta, S.; Jindal, P. Comparative evaluation of apical extrusion of debris and irrigant with three rotary instruments using crown down technique—An in vitro study. *J. Oral Biol. Craniofacial Res.* **2012**, *2*, 105–109. [CrossRef] [PubMed]
20. El Hachem, R.; Khalil, I.; Le Brun, G.; Pellen, F.; Le Jeune, B.; Daou, M.; El Osta, N.; Naaman, A.; Abboud, M. Dentinal tubule penetration of AH Plus, BC Sealer and a novel tricalcium silicate sealer: A confocal laser scanning microscopy study. *Clin. Oral Investig.* **2018**, *23*, 1871–1876. [CrossRef]
21. Hovland, E.J.; Dumsha, T.C. Leakage evaluation in vitro of the root canal sealer cement Sealapex. *Int. Endod. J.* **1985**, *18*, 179–182. [CrossRef]
22. Loushine, B.A.; Bryan, T.E.; Looney, S.W.; Gillen, B.M.; Loushine, R.J.; Weller, R.N.; Pashley, D.H.; Tay, F.R. Setting Properties and Cytotoxicity Evaluation of a Premixed Bioceramic Root Canal Sealer. *J. Endod.* **2011**, *37*, 673–677. [CrossRef]
23. Dobrzańska, J.; Dobrzański, L.B.; Gołombek, K.; Dobrzański, L.A.; Dobrzańska-Danikiewicz, A.D. Virtual approach to the comparative analysis of biomaterials used in endodontic treatment. *Processes* **2021**, *9*, 926. [CrossRef]

24. Ulusoy, Ö.I.A.; Nayr, Y.; Darendeliler-Yaman, S. Effect of different root canal sealers on fracture strength of simulated immature roots. *Oral Surg. Oral Med. Oral Pathol. Oral Radiol. Endodontology* **2011**, *112*, 544–547. [CrossRef] [PubMed]
25. Vilanova, W.V.; Carvalho-Junior, J.R.; Alfredo, E.; Sousa-Neto, M.D.; Silva-Sousa, Y.T.C. Effect intracanal irrigants on the bond strength of epoxy resin-based and methacrylate resin-based sealers to root canal walls. *Int. Endod. J.* **2012**, *45*, 42–48. [CrossRef]
26. Tay, F.R.; Pashley, D.H. Monoblocks in root canals: A hypothetical or a tangible goal. *J. Endod.* **2007**, *33*, 391–398. [CrossRef]
27. Ali, M.R.W.; Mustafa, M.; Bårdsen, A.; Bletsa, A. Fracture resistance of simulated immature teeth treated with a regenerative endodontic protocol. *Acta Biomater. Odontol. Scand.* **2019**, *5*, 30–37. [CrossRef] [PubMed]
28. Johnson, M.E.; Stewart, G.P.; Nielsen, C.J.; Hatton, J.F. Evaluation of root reinforcement of endodontically treated teeth. *Oral Surg. Oral Med. Oral Pathol. Oral Radiol. Endodontology* **2000**, *90*, 360–364. [CrossRef]
29. Gade, V.J.; Belsare, L.D.; Patil, S.; Bhede, R.; Gade, J.R. Evaluation of push-out bond strength of endosequence BC sealer with lateral condensation and thermoplasticized technique: An in vitro study. *J. Conserv. Dent.* **2015**, *18*, 124–127. [CrossRef]
30. Yendrembam; Mittal, A.; Sharma, N.; Dhaundiyal, A.; Kumari, S.; Abraham, A. Relative assessment of fracture resistance of endodontically treated teeth with epoxy resin-based sealers, AH Plus, MTA Fillapex, and Bioceramic Sealer: An In vitro study. *Indian J. Dent. Sci.* **2019**, *11*, 46. [CrossRef]
31. Saw, L.H.; Messer, H.H. Root strains associated with different obturation techniques. *J. Endod.* **1995**, *21*, 314–320. [CrossRef]
32. Shaheen, N.A.; Farag, A.M.; Alhadainy, H.A.; Darrag, A.M. Fracture resistance of endodontically treated roots using different preparation–obturation combinations. *Tanta Dent. J.* **2013**, *10*, 97–102. [CrossRef]
33. Lertchirakarn, V.; Palamara, J.E.A.; Messer, H.H. Load and strain during lateral condensation and vertical root fracture. *J. Endod.* **1999**, *25*, 99–104. [CrossRef]

applied sciences

MDPI

Article

Association of Self-Rated Pain with Clinical Peri-Implant Parameters and Cytokine Profile in Smokers and Never Smokers with and without Peri-Implantitis

Faris A. Alshahrani [1], Firas K. Alqarawi [1], Montaser N. Alqutub [2], Abdulrahman M. AlMubarak [2], Eman M. AlHamdan [3], Samar Al-Saleh [3], Paras Ahmad [4], Fahim Vohra [3,5] and Tariq Abduljabbar [3,5,*]

1 Department of Substitutive Dental Sciences, College of Dentistry, Imam Abdulrahman Bin Faisal University, Dammam 34212, Saudi Arabia; faalshahrani@iau.edu.sa (F.A.A.); fkalqarawi@iau.edu.sa (F.K.A.)
2 Department of Periodontics and Community Dentistry, College of Dentistry, King Saud Uiversity, Riyadh 11451, Saudi Arabia; alqutub@hotmail.com (M.N.A.); amalmubarak@ksu.edu.sa (A.M.A.)
3 Department of Prosthetic Dental Science, College of Dentistry, King Saud University, Riyadh 11545, Saudi Arabia; ealhamdan@ksu.edu.sa (E.M.A.); salsaleh@ksu.edu.sa (S.A.-S.); fvohra@ksu.edu.sa (F.V.)
4 Department of Oral Medicine, Dental Section, Akhtar Saeed Medical and Dental College, Lahore 53720, Pakistan; dparas2017@gmail.com
5 Chair for Biological research in Dental Health, College of Dentistry, King Saud University, Riyadh 11545, Saudi Arabia
* Correspondence: tajabbar@ksu.edu.sa; Tel.: +0096-6532-911056

Citation: Alshahrani, F.A.; Alqarawi, F.K.; Alqutub, M.N.; AlMubarak, A.M.; AlHamdan, E.M.; Al-Saleh, S.; Ahmad, P.; Vohra, F.; Abduljabbar, T. Association of Self-Rated Pain with Clinical Peri-Implant Parameters and Cytokine Profile in Smokers and Never Smokers with and without Peri-Implantitis. *Appl. Sci.* **2021**, *11*, 5559. https://doi.org/10.3390/app11125559

Academic Editor: Mary Anne Melo

Received: 6 May 2021
Accepted: 26 May 2021
Published: 16 June 2021

Abstract: To examine the association between self-perceived pain (SPP), clinical and radiographic peri-implant parameters, and biomarker levels among smokers and never smokers with and without peri-implantitis. Sixty individuals (20 smokers with peri-implantitis [group-1], 20 never smokers with peri-implantitis [group-2] and 20 never smokers without peri-implantitis [control-group]) were included. SPP was evaluated using a numeric pain rating scale (NPRS). Peri-implant plaque index (PI), probing depth (PD), and bleeding on probing (BOP) were recorded. After obtaining the samples, the levels of TNF-α, MMP-1, and IL-8 were measured. The mean SPP score in group-1, group-2, and the control group was 1.3 ± 1, 3.4 ± 1, and zero, respectively. The peri-implant mean PD ($p < 0.05$), BOP ($p < 0.05$), PI ($p < 0.05$), and crestal bone loss (CBL) ($p < 0.05$) were significantly higher among test groups than the control group. The levels of TNF-α, MMP-1, and IL-8 were significantly raised among group-1 and group-2 than the control group. A significant correlation between increasing SPP and PICF TNF-α, MMP-1, and IL-8 levels was observed based on regression analysis. Proinflammatory biomarkers were higher in smokers with peri-implantitis than never smokers with and without peri-implantitis, with a significant association between the proinflammatory cytokines and SPP.

Keywords: cigarette smoking; cytokines; pain; peri-implantitis; peri-implant crevicular fluid; inflammation

1. Introduction

Pain is an unpleasant sensory and emotional experience associated with actual or potential tissue damage or described in terms of such damage [1]. It can be associated with inflammation of soft tissue surrounding the dental implant and crestal bone loss (CBL); also called peri-implantitis [2]. A systematic review reported that the evaluation of self-perceived pain (SPP) is also a diagnostic parameter of peri-implantitis besides other diagnostic criteria such as CBL, bleeding on probing (BOP), probing depth (PD), and suppuration [3]. Furthermore, the International Congress of Implantologists (ICOI) also suggested SPP to be utilized as a diagnostic parameter for implant failure and peri-implantitis [4] The numeric pain rating scale (NPRS), developed by Downie et al. [5], is a valid and reliable scale often utilized to examine the SPP in subjects with musculoskeletal diseases, for instance, osteoarthritis [6]. The NPRS has also been utilized to evaluate links with dental implant surgeries [7].

The abutment screw loosening or failure is a rare, but very unpleasant failure. Several studies have reported that, after dental implant osseointegration, abutment screw loosening seems to be the most frequent issue linked with implants [8–11]. Many factors contribute to the etiology of these technical failures, such as fabrication failures, non-passive fit of the suprastructures, improper placement technique, fatigue, excessive occlusal load, and the utilization of unfavorable components including copy products [12]. A few reports suggest that most of the failures are related to the suprastructure rather than to the dental implants themselves [13].

Cigarette smoking is a well-established predisposing factor for both peri-implant and periodontal soft tissue inflammation; [14] and CBL surrounding dental implants and natural teeth [15]. One justification in this aspect might be that the generation and deposition of advanced glycation end products (AGEs) in the periodontal tissues is increased by cigarette smoking [16]. The linkage of AGEs with their receptors (RAGE), inflammatory cytokines including matrix metalloproteinase (MMP)-1, interleukin (IL)-8, and tumor necrosis factor-alpha (TNF-α) are generated by human gingival fibroblasts that stimulate inflammation [17]. Furthermore, collagen degradation is increased by cigarette smoking via influencing tissue inhibitors of metalloproteinases [18]. Moreover, the RAGE's expression in the gingival tissues is upregulated; the production of reactive oxygen species (ROS) is stimulated by metabolites of nicotine, i.e., nornicotine which damages the periodontal tissues. Besides contributing to delayed periodontal wound healing, these variables might also compromise peri-implant tissue healing. The destructive inflammatory cytokines cause deteriorating clinical peri-implant parameters including plaque index (PI), PD, gingival index (GI), elevated CBL on the distal and mesial aspects of the dental implant. If untreated, implant failure may be caused [19]. Alqahtani et al. [20] evaluated the clinical peri-implant parameters and levels of cotinine in peri-implant crevicular fluid (PICF) of thirty-five never-smokers and thirty-five tobacco smokers. The findings reported that cotinine levels and peri-implantitis in the PICF were considerably lower in never smokers as compared to tobacco smokers. Furthermore, studies have suggested that nicotine harms human periodontal ligament cells and gingival fibroblasts by increasing the generation of inflammatory mediators and compromising cellular proliferation [21,22]. Regardless of its harmful effects on peri-implant and periodontal health, cigarette smoking is not an absolute contraindication to dental implant treatment [23].

To date, no study has reported the association between PICF levels of TNF-α, MMP-1, and IL-8 with clinic-radiographic peri-implant parameters (PI, PD, BOP, and CBL) along with SPP among patients with peri-implantitis. In the current study, it is hypothesized that clinical and radiographic peri-implant parameters along with pain scores are worse and the levels of TNF-α, MMP-1, and IL-8 in the PICF are higher in cigarette smokers with peri-implantitis in comparison with never-smokers with peri-implantitis. The present study aimed to assess the association between SPP, clinical and radiographic peri-implant parameters, and levels of PICF TNF-α, MMP-1, and IL-8 among cigarette smokers and never smokers with and without peri-implantitis.

2. Materials and Methods

2.1. Ethical Considerations

This study was submitted, reviewed, and approved by King Saud University, Riyadh, Saudi Arabia (UDRC/019-12). After reading the informed consent document, written in simple Arabic and English, all the volunteering participants provided written informed consent. The ethical standards of the 1964 Helsinki declaration and national and/or institutional research committee were strictly followed while performing all the procedures. All the participants were aware that they had the right to withdraw at any time throughout the study course without any indirect or direct consequences.

2.2. Inclusion and Exclusion Criteria

The inclusion criteria comprised of patients with the following characteristics: (1) reading and signing the written informed consent document; (2) patients aged $\geq$ 25 years who underwent dental implant treatment; (3) individuals suffering from peri-implantitis, i.e., peri-implant BOP at $\geq$30% sites, PD $\geq$ 4 mm, and CBL $\geq$ 3 mm [24]; (4) individuals without peri-implantitis, i.e., BOP of <30% sites, PD of <4 mm, and CBL of <3 mm [24]; (5) self-reported tobacco consumers: subjects who have been smoking a minimum of one cigarette/day over the past 2 years [25]; (6) never smokers: persons that stated to have never utilized tobacco in any form [26].

The exclusion criteria consisted of individuals with the following characteristics: (1) individuals who refused to participate and/or sign the written consent document; (2) individuals consuming smokeless tobacco products and/or areca nut; (3) dual smoker, i.e., persons smoking tobacco along with other nicotinic products including waterpipe, bidi, and electronic cigarettes; (4) individuals who consumed alcohol habitually; (5) completely edentulous individuals; (6) individuals with maligned dentition; (7) patients suffering from medical conditions such as diabetes mellitus, renal and hepatic disorders, cardiovascular diseases, and patients suffering from the HIV infection; (8) gestation and/or nursing individuals; (9) individuals who reported to have utilized steroids, NSAIDs, probiotics, antibiotics and bisphosphonates over the past 3 months.

2.3. Study Design and Participants

This is a retrospective case-control study conducted between January 2019 and February 2020 at the College of Dentistry, King Saud University, Riyadh, Saudi Arabia. The participants were classified into three groups on the basis of peri-implant status and self-reported habits of cigarette smoking: (1) cigarette smokers with peri-implantitis (Group-1); (2) never smokers with peri-implantitis (Group-2); and (3) never smokers without peri-implantitis (control-group).

2.4. Questionnaire

Information related to age, sex, highest education level, family history of smoking, daily frequency and duration of cigarette smoking, daily flossing and tooth brushing, duration of functional dental implants, location, and the number of dental implants in either jaw were collected. The questionnaire also collected data relating to SPP utilizing the NPRS. It was suggested that participants choose a number ranging from 0 to 10 that most appropriately fit their perceived pain's intensity with 10 and 0 depicting the worst pain possible and no pain, respectively [27]. A trained investigator (T.A.) distributed the questionnaire to all participants.

2.5. Parameters Related to Dental Implant

A calibrated investigator (T.A.) gathered the data from patient's records relating to the implant dimensions (diameter x length), jaw location (mandible or maxilla), surface characteristics (smooth or moderately rough surfaces), insertion torque (in Newton centimeters [Ncm]), implant–abutment connection (platform-switching), prosthetic restoration type (screw- or cement-retained). "Implant success" was described on the basis of the following characteristics: (1) existence of a fully functional implant within the oral cavity; (2) self-evaluated satisfaction of the patient regarding implant and the associated prosthesis [28].

2.6. Evaluation of Peri-Implant Clinic-Radiographic Parameters

As explained elsewhere [20], a trained investigator (F.A.V.) calculated and reported the peri-implant PD, BOP, PI, and CBL (distal and mesial) around every dental implant, as recommended by the specific guidelines declared in the consensus report of the Eleventh European Workshop on Periodontology [29]. To summarize, the peri-implant PD, BOP, and PI were calculated at 6 sites per implant, i.e., disto-palatal/-lingual, mid-palatal/-lingual, mesio-palatal/-lingual, distobuccal, mid-buccal, mesiobuccal. Utilizing a graded probe

(Hu-Friedy, Chicago, IL, USA), the measurement of peri-implant PD was carried out to the nearest mm. The digital bitewing radiographs, which were standardized utilizing long cone parallel methods, were employed to record the CBL [30]. The recording of the linear distance was performed 0.2 cm below the abutment–implant junction to the alveolar crest. The CBL was defined as the bone loss median measured on mesial and distal aspects of the dental implant.

2.7. Collection of PICF

PICF collection was carried out as described elsewhere [31]. The mechanical cleaning of the supragingival plaque present on the implant-supported crown was carried out. After the careful isolation of the peri-implant sites using sterile cotton, an air syringe was used for drying purposes. By introducing 1–2 mm of paper points (Periopaper, Pro Flow, Amityville, NY, USA) into the peri-implant sulcus for 30 s, the collection of PICF samples was performed. After discarding the samples contaminated with bacteria, blood, or saliva, the collection of the new specimens was carried out from the same site. An electronic calibrated gingival fluid device (Periotron 6000, Amityville, NY, USA) was utilized to measure the fluid sample volume. The elution of the collected PICF was carried out with 1 mL of phosphate-buffered saline (PBS) prior to freezing at −80 °C.

2.8. Evaluation of PICF Levels of TNF-α, MMP-1, and IL-8

Centrifugation of PICF samples was carried out at 15,000× g for 15 min. Quantification of the levels of biomarkers was performed using enzyme-linked immunosorbent assay (ELISA) according to the manufacturer's guidelines (Quantikine, R&D Systems, Minneapolis, MN, USA) and calculated as pg/mL. A trained investigator (P.A.) carried out all laboratory-based analysis, (i.e., *kappa* = 0.78). In brief, the addition of 100 mL samples and standards were incorporated into the corresponding wells. Then, their incubation was performed overnight at 4 °C after covering them properly. Later, to discard the solution, a multi-channel pipette was utilized and cleaned four times using a 0.3 mL 1× wash solution. Post final wash, the removal of the residual liquid was carried out via aspiration. Then, the inversion of the plate was performed and blotted with sterile paper towels. Inclusion of 0.1 mL of 1× Biotinylated anti-Human CRP Detector Antibody to individual wells was carried out. Later, this was incubated at room temperature, (i.e., 25 °C) for 60 min. The rewashing of the solution was carried out after disposing it off. Before incubating it at 25 °C for 45 min, 0.1 mL of 1× HRP-Streptavidin solution was incorporated into each individual well. The solution was re-discarded and washed again. To each well, 0.1 mL of TMB One-Step Substrate Reagent was incorporated and incubation was carried out at 25 °C for half an hour in a dark room. The addition of 50 μL of stop solution to each well was performed before instantly taking the reading at 450 nm. Prior to each individual assay, standard curves were produced and employed for plotting outcomes. For the levels of PICF of TNF-α, MMP-1, and IL-8, the ELISA's sensitivity was 96%, 97.2%, and 96.8%, respectively.

2.9. Statistical and Power Analysis

The Statistical Package for Social Sciences (SPSS) version 21.0 was used to perform statistical analysis. The goodness of fit test was carried out to assess the normality of the data. Unless otherwise specified, the results were shown as mean ± SD of the mean. To assess between-group differences of non-normal data distribution, the Kruskal–Wallis nonparametric test was conducted. Bivariate multiple regression analysis was carried out to evaluate the association between PICF cytokine levels and SPP. A total of 15 patients were required with 25 dental implants per group to attain a power of 90% with a probing depth of 1 mm. To allow a 20% attrition for compensation, a total of 20 patients were selected. A p-value of <0.05 was considered statistically significant.

3. Results

3.1. General Description of Study Participants

A total of 60 patients (29 males and 31 females) participated in this study and signed the written informed consent and completed the clinical trial. Group-1 and group-2 participants had a mean age of 41.7 ± 6.6 and 39.4 ± 5.8 years, respectively, while the control group had a mean age of 38.2 ± 7.3 years. Among the smokers and never smokers with peri-implantitis, 42 and 36 dental implants were examined, respectively, while 28 dental implants were evaluated among never smokers without peri-implantitis. Among group-1 and group-2, an average of dental implants in service was 37.4 ± 3.2 and 31.6 ± 2.1 months, respectively, while among the control group an average of dental implants in use was 29.9 ± 5.0 months (Table 1).

Table 1. Demographics of the study groups.

Characteristics	Group-I	Group-II	Group-III
Total number of samples '*n*'	20	20	20
Mean age (±SD)	41.7 ± 6.6	39.4 ± 5.8	38.2 ± 7.3
Gender (M/F)	11/9	8/12	10/10
Number of implants assessed	42	36	28
Duration of implants in months (±SD)	37.4 ± 3.2	31.6 ± 2.1	29.9 ± 5.0

3.2. Clinical and Radiographic Peri-Implant Parameters

The peri-implant mean PD ($p < 0.05$) and mesial ($p < 0.05$) and distal ($p < 0.05$) CBL were significantly higher among group-1 than group-2 and control group. However, the peri-implant mean BOP ($p < 0.05$) and PI ($p < 0.05$) were significantly higher in group-2 when compared with group-1 and the control group. Overall, the peri-implant mean PI ($p < 0.05$), BOP ($p < 0.05$), PD ($p < 0.05$), and CBL ($p < 0.05$) were significantly higher among test groups than the control group. The mean SPP score was highest in group-2 (3.4 ± 1) when compared with group-1 (1.3 ± 1) and the control group (0 ± 0), respectively (Table 2).

Table 2. Peri-implant clinical parameters of the research groups.

Variables (Mean ± SD)	Group-I (n = 20)	Group-II (n = 20)	Group-III (n = 20)
PI in % (±SD)	41.5 ± 8.3 ^{a}	46.6 ± 11.4 ^{a}	13.7 ± 5.1 ^{b}
BOP in % (±SD)	21.6 ± 6.9 ^{c}	53.5 ± 7.7 ^{d}	15.8 ± 7.0 ^{e}
PD in mm (±SD)	5.0 ± 0.5 ^{f}	4.8 ± 0.3 ^{f}	1.2 ± 0.8 ^{g}
CBL in mm (±SD)	2.9 ± 0.6 ^{h}	2.7 ± 0.4 ^{h}	0.7 ± 0.02 ^{i}
SPP score	1.3 ± 1 ^{j}	3.4 ± 1 ^{k}	0 ± 0 ^{j}

Dissimilar lowercase letters indicate a statistically significant difference ($p < 0.05$) between groups along rows. Abbreviations: BOP = bleeding on probing; CBL = crestal bone loss; PD = probing depth; PI = plaque index; SPP = self-perceived pain.

3.3. Peri-Implant Crevicular Fluid Cytokine Levels and Self-Perceived Pain

PICF levels of TNF-α, MMP-1, and IL-8 were raised significantly among group-1 and group-2 than the control group (Table 3).

Table 3. Peri-implant crevicular fluid cytokine levels of the study groups.

Cytokines (pg/mL)	Group-I (n = 20)	Group-II (n = 20)	Group-III (n = 20)
TNF-α	648 ± 335 ^{a}	395 ± 182 ^{b}	87 ± 23 ^{c}
MMP-1	4612 ± 2296 ^{d}	3179 ± 1498 ^{e}	174 ± 82 ^{f}
IL-8	442 ± 202 ^{g}	387 ± 164 ^{g}	53 ± 22 ^{h}

Dissimilar lowercase letters indicate a statistically significant difference ($p < 0.05$) between groups along rows.

Statistically, a significant association was observed between SPP and TNF-α, MMP-1, and IL-8 expression in the PICF of test groups and control group as per bivariate multiple logistic regression analysis (Table 4).

Table 4. Statistical comparisons of the SPP score with PICF cytokines levels.

SPP Score	OR	95% CI	*p*-Value
TNF-α			
1	0.45	0.32–0.58	
2	0.13	0.09–0.17	
3	0.72	0.62–0.82	0.032
4	0.32	0.15–0.49	
5	0.28	0.22–0.34	
6	0.16	0.10–0.22	
MMP-1			
1	0.15	0.05–0.25	
2	0.36	0.31–0.41	
3	0.52	0.43–0.61	
4	0.23	0.15–0.46	0.044
5	0.19	0.11–0.27	
6	0.08	0.05–0.11	
IL-8			
1	0.16	0.12–0.21	
2	0.47	0.40–0.54	
3	0.39	0.21–0.57	
4	0.61	0.56–0.66	0.0082
5	0.07	0.06–0.08	
6	0.26	0.18–0.34	

Abbreviations: CI = confidence interval; IL-8 = interleukin-8; MMP-1 = matrix metalloproteinase-1; OR = odds ratio; PICF = peri-implant crevicular fluid; SPP = self-perceived pain; TNF-α = tumor necrosis factor-alpha.

4. Discussion

It is postulated that SPP, peri-implant clinical, and radiographic parameters are worse among cigarette smokers with peri-implantitis than never smokers without peri-implantitis. In this study, the clinic-radiographic peri-implant parameters were poorer among cigarette smokers with peri-implantitis than never smokers without peri-implantitis; no statistically significant correlation was observed between SPP and expression of TNF-α, MMP-1, and IL-8. However, several variables might have altered the findings, therefore, these results should be interpreted with extreme caution.

A unique ecological niche is formed by the gingival sulcus for bacterial colonization. The bacterial biofilm in the gingival sulcus promotes inflammation in the surrounding connective tissues [32]. An inflammatory process is triggered by the imbalance between host response and microbial challenge at the implant–soft tissue interface [33]. Cytokines, including IL-6, IL-1B, TNF-α, are released from the cells of neutrophils, macrophages, connective tissue fibroblasts, dendritic cells, and gingival epithelium. Moreover, a number of enzymes, including aspartate aminotransferase, alkaline phosphatase, and MMPs, are formed by osteoclasts, fibroblasts, and neutrophils, resulting in the degradation of connective tissue collagen and alveolar bone [34,35]. Over 90 different molecular components in gingival crevicular fluid have been assessed for potential periodontal disease diagnosis linked with the natural dentition [36]. However, considerably fewer PICF components have been investigated around dental implants.

In the current study, certain disparities were also observed. For example, clinical peri-implant examination revealed that PI and BOP were significantly raised in never smokers with peri-implantitis than smokers with peri-implantitis and the control group. Moreover, in cigarette smokers, BOP is masked as nicotine causes the vasoconstriction of gingival blood vessels [37]. From a clinical perspective, the findings of this study conform to the proposed hypothesis as scores of peri-implant PD are significantly elevated among cigarette smokers than never smokers. One explanation for this finding is that cell death and formation of matrix metalloproteinases are induced and enhanced by cigarette smoke extracts, respectively, leading to the deterioration of the extracellular matrix proteins such as collagen [38]. Furthermore, the expression for RAGE is increased four times by nicotine in human gingival cells [39]; an increased AGE/RAGE interaction aggravates periodontal tissue inflammation [39].

In the current study, smokers and never smokers with peri-implantitis did not show a significant difference in CBL. Among all three groups, the subjects had a mean age between 38 and 42 years. It is well-established that advancing or increasing age is a predisposing factor for the increased loss of alveolar bone around dental implants and natural dentition [14]. In one study, Javed et al. [25] reported that alveolar bone loss around teeth was significantly less among younger non-smokers ($\leq$45 years of age) than old non-diabetic smokers ($\geq$65 years of age). Furthermore, subjects in group-1 (smokers with peri-implantitis) had a cigarette smoking history of one cigarette per day over the past two years. It is possible that participants with such a brief history of tobacco smoking may exhibit better radiographic bone levels when compared with subjects with a prolonged history of cigarette smoking, (i.e., >15 packs per year) [14]. However, further research is required for testing these hypotheses.

The current study reports that TNF-α, MMP-1, and IL-8 levels were significantly raised among smokers and never smokers with peri-implantitis than never smokers without peri-implantitis. TNF-α, MMP-1, and IL-8 appear to have a crucial role in the destruction of the peri-implant tissue. TNF-α is a proinflammatory mediator and its level reflects the bacterial amount and the levels of inflammation [40]. MMPs are considered to be involved in wound repair, normal tissue turnover, and periodontal destruction; MMP-1 may initiate the deterioration of the extracellular matrix. Overproduction of MMP-1 may cause accelerated degradation of the matrix in pathologic conditions including peri-implantitis [41]. IL-8 increase causes the activation of polymorphonuclear cells and influences their selective migration from gingival blood vessels. This leads to an increased amount of cells in a limited time and produces inflammatory conditions. Hence, an excessive quantity of IL-8 might indicate the incipient stage of peri-implantitis [42]. Several studies have reported that the secretion of TNF-α, MMP-1, and IL-8 is upregulated by nicotine in tobacco; which appears to play a vital role in the destruction of alveolar bone around dental implants and natural dentition [17]. Elevated levels of these proinflammatory cytokines have been observed in the PICF of patients with peri-implantitis [43]. Moreover, a meta-analysis reported that cigarette smoking jeopardizes bone-to-implant contact by impairing new bone formation around the dental implant [44]. Another explanation in this regard is that cigarette smoking escalates the production and accumulation of AGEs in periodontal tissues [39]. Strong interfaces between AGEs and RAGEs have been linked with the production of ROS that promotes oxidative burst within gingival tissues, functional changes of phagocytosis and chemotaxis of polymorphonuclear cells (PMNs), alleviated formation of antibodies, increased attachment of bacterial adhesion, and raised local and systemic burden by escalating the expression of cytokines in the crevicular fluid and serum [45]. These mechanisms have been associated with the inflammation of oral connective tissues and destruction of alveolar bone around dental implants and natural teeth in cigarette smokers [16].

In literature, a dearth of reports exist that assessed SPP and its association with the inflammatory cytokines' levels discharged saliva, serum, or PICF. In this aspect, the authors hypothesized that PICF levels of TNF-α, MMP-1, IL-8 are raised in smokers with peri-implantitis with high SPP scores than non-smokers with and without peri-implantitis. The findings of the present study are in opposition with this hypothesis as no statistically significant association between SPP and PICF levels of TNF-α, MMP-1, IL-8. One justification in this regard might be that an array of factors such as coping strategies, cognition, expectations, beliefs, and interpretation influence pain evaluation.

Furthermore, from these findings, it is not possible to calculate the minimum amount of proinflammatory cytokines required to evoke SPP in individuals with peri-implantitis. Despite a well-documented association between pain conditions and cytokines [46,47], the association between pain perception and cytokine expression in oral fluids remains unclear and poorly understood. Hence, further research is required in this regard.

Although questionnaires are not particularly reliable regarding clinical peri-implant parameters, including PI, PD, and BOP [48], they are valid and reliable tools for evaluation of SPP and other oral symptoms [48]. Moreover, validated and well-designed questionnaires could too yield valuable data in survey-based epidemiological large sample-size reports [26]. In the current study, NPRS was utilized to assess the SPP following the recommendations by Downie and co-workers [5] almost forty years ago, who reported that the NPRS (0–10) was a better method to assess SPP than other scales including the four-point descriptive and the visual analogue scales. Because NPRS was the only utilized scale for assessing the SPP, it is difficult to either contradict or support the Downie study [5]. Based on the present study's findings, the NPRS appears to be an efficient scale for pain evaluation in periodontology and implant dentistry. However, future studies utilizing several pain evaluation scales are required to identify the most efficient and suitable pain scale that can be employed on individuals with peri-implant diseases. It is important to recognize the limitations of this study. First, the cross-sectional study and self-reported results that depend on the subject's recall capacities might have influenced the outcomes of our findings. The assessment of pain biomarker(s) would have helped to evaluate the pain parameter on the molecular level. Second, strict criteria were adopted regarding patient selection. For example, patients with endocrine disorders and tobacco-product consumers were excluded. As habitually consuming tobacco products is a predisposing factor of peri-implant diseases [49], it was speculated that the levels of PICF proinflammatory cytokines and the severity of SPP are less in never smokers than smokers with peri-implantitis. The primary strength of this study is the assessment of local proinflammatory cytokines. In terms of accuracy, the use of bitewing radiographs in this study provided better-standardized measurements. Future research studies using several pain evaluation scales and biomarkers related to pain and dental implant therapy are required to identify the most efficient and suitable diagnostic criterion that could be employed on individuals with peri-implant diseases. In addition, longitudinal reports might help to evaluate the PICF biomarkers' evolution over time for better understanding in terms of peri-implant infections among smokers and non-smokers.

5. Conclusions

Within the limitations of the present study, proinflammatory biomarkers were higher in smokers with peri-implantitis than never smokers with and without peri-implantitis, with a significant association between the proinflammatory cytokines and self-perceived pain.

Author Contributions: Conceptualization, F.A.A., F.K.A. and M.N.A.; methodology, F.A.A., F.K.A., E.M.A., T.A. and M.N.A. software.; validation, A.M.A., S.A.-S., F.V., M.N.A. and T.A. formal analysis, M.N.A., E.M.A., S.A.-S. and F.K.A.; investigation, F.A.A., F.K.A., A.M.A., S.A.-S. and F.V.; resources, S.A.-S., E.M.A.; data curation, M.N.A., A.M.A. and F.A.A.; writing—original draft preparation, F.V., S.A.-S. and P.A.; writing—review and editing, F.A.A., A.M.A. and F.K.A.; supervision, F.V. All authors have read and agreed to the published version of the manuscript.

Funding: The authors are grateful to the Deanship of Scientific Research, King Saud University for funding through Vice Deanship of Scientific Research Chairs, Research Chair for Biological Research in Dental Health.

Institutional Review Board Statement: This study was submitted, reviewed, and approved by King Saud University, Riyadh, Saudi Arabia (UDRC/019-12). The ethical standards of the 1964 Helsinki declaration and national and/or institutional research committee were strictly followed while performing all the procedures. Informed written consent was obtained from each subject before conducting any procedures. Additional information on the study was provided verbally by the study investigator or in a written format.

Informed Consent Statement: Not applicable.

Data Availability Statement: The data is available on contact from the corresponding author.

Acknowledgments: The authors are grateful to the Deanship of Scientific Research, King Saud University for funding through Vice Deanship of Scientific Research Chairs, Research Chair for Biological Research in Dental Health.

Conflicts of Interest: The authors declare no conflict of interest.

References

1. Nugraha, B.; Gutenbrunner, C.; Barke, A.; Karst, M.; Schiller, J.; Schäfer, P.; Falter, S.; Korwisi, B.; Rief, W.; Treede, R.D.; et al. The IASP classification of chronic pain for ICD-11: Functioning properties of chronic pain. *Pain* **2019**, *160*, 88–94. [CrossRef] [PubMed]
2. Coli, P.; Christiaens, V.; Sennerby, L.; Bruyn, H.D. Reliability of periodontal diagnostic tools for monitoring peri-implant health and disease. *Periodontology 2000* **2017**, *73*, 203–217. [CrossRef]
3. Ramanauskaite, A.; Juodzbalys, G. Diagnostic principles of peri-implantitis: A systematic review and guidelines for peri-implantitis diagnosis proposal. *J. Oral Maxillofac. Res.* **2016**, *7*, e8. [CrossRef]
4. Misch, C.E.; Perel, M.L.; Wang, H.L.; Sammartino, G.; Galindo-Moreno, P.; Trisi, P.; Steigmann, M.; Rebaudi, A.; Palti, A.; Pikos, M.A.; et al. Implant success, survival, and failure: The International Congress of Oral Implantologists (ICOI) pisa consensus conference. *Implant Dent.* **2008**, *17*, 5–15. [CrossRef]
5. Downie, W.; Leatham, P.; Rhind, V.; Wright, V.; Branco, J.; Anderson, J. Studies with pain rating scales. *Ann. Rheum. Dis.* **1978**, *37*, 378–381. [CrossRef]
6. Saltychev, M.; Bärlund, E.; Laimi, K. Correlation between the pain numeric rating scale and the 12-item WHO Disability Assessment Schedule 2.0 in patients with musculoskeletal pain. *Int. J. Rehabil. Res.* **2018**, *41*, 87–91. [CrossRef]
7. Iero, P.T.; Mulherin, D.R.; Jensen, O.; Berry, T.; Danesi, H.; Razook, S.J. A prospective, randomized, open-Label study comparing an Opioid-Sparing Postsurgical Pain Management Protocol With and Without Liposomal Bupivacaine for Full-Arch Implant Surgery. *Int. J. Oral Maxillofac. Implant.* **2018**, *33*, 1155–1164. [CrossRef] [PubMed]
8. Scarano, A.; Murmura, G.; Sinjiari, B.; Sollazzo, V.; Spinelli, G.; Carinci, F. Analysis and structural examination of screw loosening in oral implants. *Int. J. Immunopathol. Pharmacol.* **2011**, *24*, 77–81. [CrossRef] [PubMed]
9. Assenza, B.; Scarano, A.; Leghissa, G.; Carusi, G.; Thams, U.; Roman, F.S.; Piattelli, A. Screw-vs Cement-implant–retained Restorations: An Experimental Study in the Beagle. Part 1. Screw and Abutment Loosening. *J. Oral Implantol.* **2005**, *31*, 242–246. [CrossRef]
10. Jemt, T.; Laney, W.R.; Harris, D.; Henry, P.J.; Krogh, P.H., Jr.; Polizzi, G.; Zarb, G.A.; Herrmann, I. Osseointegrated implants for single tooth replacement: A 1-year report from a multicenter prospective study. *Int. J. Oral Maxillofac. Implant.* **1991**, *6*, 29–36.
11. Al-Aali, K.A.; ArRejaie, A.S.; Alrahlah, A.; AlFawaz, Y.F.; Abduljabbar, T.; Vohra, F. Clinical and radiographic peri-implant health status around narrow diameter implant-supported single and splinted crowns. *Clin. Implant Dent. Relat. Res.* **2019**, *21*, 386–390. [CrossRef]
12. Cranin, A.N.; Dibling, J.B.; Simons, A.; Klein, M.; Sirakian, A. Report of the incidence of implant insert fracture and repair of Core-Vent dental implants. *J. Oral Implantol.* **1990**, *16*, 184–188.
13. Dixon, D.L.; Breeding, L.C.; Sadler, J.P.; McKay, M.L. Comparison of screw loosening, rotation, and deflection among three implant designs. *J. Prosthet. Dent.* **1995**, *74*, 270–278. [CrossRef]
14. Alasqah, M.N.; Alfawaz, Y.F.; Aldahiyan, N.; Vohra, F.; Alotaibi, B.M.; Abduljabbar, T. Longitudinal assessment of clinical and radiographic periimplant status around narrow and regular diameter implants placed in cigarette-smokers and nonsmokers. *Clin. Implant Dent. Relat. Res.* **2019**, *21*, 910–915. [CrossRef]
15. Veitz-Keenan, A. Marginal bone loss and dental implant failure may be increased in smokers. *Evid. Based Dent.* **2016**, *17*, 6–7. [CrossRef] [PubMed]
16. Akram, Z.; Alqahtani, F.; Alqahtani, M.; Al-Kheraif, A.A.; Javed, F. Levels of advanced glycation end products in gingival crevicular fluid of chronic periodontitis patients with and without type-2 diabetes mellitus. *J. Periodontol.* **2020**, *91*, 396–402. [CrossRef]
17. Yu, S.; Li, H.; Ma, Y.; Fu, Y. Matrix metalloproteinase-1 of gingival fibroblasts influenced by advanced glycation end products (AGEs) and their association with receptor for AGEs and nuclear factor-κB in gingival connective tissue. *J. Periodontol.* **2012**, *83*, 119–126. [CrossRef] [PubMed]
18. Zhang, W.; Fang, M.; Song, F.; Windsor, L.J. Effects of cigarette smoke condensate and nicotine on human gingival fibroblast-mediated collagen degradation. *J. Periodontol.* **2011**, *82*, 1071–1079. [CrossRef]
19. Al-Aali, K.A.; AlHelal, A.; Alhamoudi, N.; Alhenaki, A.M.; Javed, F.; Abduljabbar, T. Assessment of advanced glycation end products in the peri-implant sulcular fluid among moderate cigarette-smokers and nonsmokers with peri-implantitis. *Clin. Implant Dent. Relat. Res.* **2020**, *22*, 1–7. [CrossRef]
20. Alqahtani, F.; Alqahtani, M.; Albaqawi, A.H.; Al-Kheraif, A.A.; Javed, F. Comparison of cotinine levels in the peri-implant sulcular fluid among cigarette and waterpipe smokers, electronic-cigarette users, and nonsmokers. *Clin. Implant Dent. Relat. Res.* **2019**, *21*, 702–707. [CrossRef] [PubMed]

21. Holliday, R.S.; Campbell, J.; Preshaw, P.M. Effect of nicotine on human gingival, periodontal ligament and oral epithelial cells. A systematic review of the literature. *J. Dent.* **2019**, *86*, 81–88. [CrossRef]
22. Alamri, A.; Semlali, A.; Jacques, E.; Alanazi, M.; Zakrzewski, A.; Chmielewski, W.; Rouabhia, M. Long-term exposure of human gingival fibroblasts to cigarette smoke condensate reduces cell growth by modulating Bax, caspase-3 and p53 expression. *J. Periodontal Res.* **2015**, *50*, 423–433. [CrossRef] [PubMed]
23. DeLuca, S.; Habsha, E.; Zarb, G.A. The effect of smoking on osseointegrated dental implants. Part I: Implant survival. *Int. J. Prosthodont.* **2006**, *19*, 491–498.
24. Abduljabbar, T.; Vohra, F.; Ullah, A.; Alhamoudi, N.; Khan, J.; Javed, F. Relationship between self-rated pain and peri-implant clinical, radiographic and whole salivary inflammatory markers among patients with and without peri-implantitis. *Clin. Implant Dent. Relat. Res.* **2019**, *21*, 1218–1224. [CrossRef]
25. Alsahhaf, A.; Alshagroud, R.S.; Al-Aali, K.A.; Alofi, R.S.; Vohra, F.; Abduljabbar, T. Survival of titanium-zirconium and titanium dental implants in cigarette-smokers and never-smokers: A 5-year follow-up. *Chin. J. Dent. Res.* **2019**, *22*, 265–272. [PubMed]
26. Vohra, F.; Bukhari, I.A.; Sheikh, S.A.; Albaijan, R.; Naseem, M. Comparison of self-rated oral symptoms and periodontal status among cigarette smokers and individuals using electronic nicotine delivery systems. *J. Am. Coll. Health* **2020**, *68*, 788–793. [CrossRef]
27. Jensen, M.P.; McFarland, C.A. Increasing the reliability and validity of pain intensity measurement in chronic pain patients. *Pain* **1993**, *55*, 195–203. [CrossRef]
28. Clark, D.; Levin, L. Dental implant management and maintenance: How to improve long-term implant success. *Quintessence Int.* **2016**, *47*, 417–423.
29. Lang, N.P.; Berglundh, T.; Working Group 4 of Seventh European Workshop on Periodontology. Periimplant diseases: Where are we now?–Consensus of the Seventh European Workshop on Periodontology. *J. Clin. Periodontol.* **2011**, *38*, 178–811. [CrossRef]
30. Khocht, A.; Janal, M.; Harasty, L.; Chang, K.M. Comparison of direct digital and conventional intraoral radiographs in detecting alveolar bone loss. *J. Am. Dent. Assoc.* **2003**, *134*, 1468–1475. [CrossRef]
31. Alasqah, M.N.; Al-Shibani, N.; Al-Aali, K.A.; Qutub, O.A.; Abduljabbar, T.; Akram, Z. Clinical indices and local levels of inflammatory biomarkers in per-implant health of obese and nonobese individuals. *Clin. Implant Dent. Relat. Res.* **2019**, *21*, 80–84. [CrossRef]
32. Costalonga, M.; Herzberg, M.C. The oral microbiome and the immunobiology of periodontal disease and caries. *Immunol. Lett.* **2014**, *162*, 22–38. [CrossRef] [PubMed]
33. Wang, H.L.; Garaicoa-Pazmino, C.; Collins, A.; Ong, H.S.; Chudri, R.; Giannobile, W.V. Protein biomarkers and microbial profiles in peri-implantitis. *Clin. Oral Implant. Res.* **2016**, *27*, 1129–1136. [CrossRef]
34. Barros, S.P.; Williams, R.; Offenbacher, S.; Morelli, T. Gingival crevicular fluid as a source of biomarkers for periodontitis. *Periodontology 2000* **2016**, *70*, 53–64. [CrossRef] [PubMed]
35. Paknejad, M.; Emtiaz, S.; Khoobyari, M.M.; Gharb, M.T.; Yazdi, M.T. Analysis of aspartate aminotransferase and alkaline phosphatase in crevicular fluid from implants with and without peri-implantitis. *Implant Dent.* **2006**, *15*, 62–69. [CrossRef]
36. Loos, B.G.; Tjoa, S. Host-derived diagnostic markers for periodontitis: Do they exist in gingival crevice fluid? *Periodontology 2000* **2005**, *39*, 53–72.
37. Clarke, N.; Shephard, B. The effects of epinephrine and nicotine on gingival blood flow in the rabbit. *Arch. Oral Biol.* **1984**, *29*, 789–793. [CrossRef]
38. Bulmanski, Z.; Brady, M.; Stoute, D.; Lallier, T.E. Cigarette smoke extract induces select matrix metalloproteinases and integrin expression in periodontal ligament fibroblasts. *J. Periodontol.* **2012**, *83*, 787–796. [CrossRef]
39. Katz, J.; Caudle, R.M.; Bhattacharyya, I.; Stewart, C.M.; Cohen, D.M. Receptor for advanced glycation end product (RAGE) upregulation in human gingival fibroblasts incubated with nornicotine. *J. Periodontol.* **2005**, *76*, 1171–1174. [CrossRef] [PubMed]
40. Al-Aali, K.A.; Alrabiah, M.; ArRejaie, A.S.; Abduljabbar, T.; Vohra, F.; Akram, Z. Peri-implant parameters, tumor necrosis factor-alpha, and interleukin-1 beta levels in vaping individuals. *Clin. Implant Dent. Relat. Res.* **2018**, *20*, 410–415. [CrossRef]
41. Peake, N.J.; Khawaja, K.; Myers, A.; Jones, D.; Cawston, T.E.; Rowan, A.D.; Foster, H.E. Levels of matrix metalloproteinase (MMP)-1 in paired sera and synovial fluids of juvenile idiopathic arthritis patients: Relationship to inflammatory activity, MMP-3 and tissue inhibitor of metalloproteinases-1 in a longitudinal study. *Rheumatology* **2005**, *44*, 1383–1389. [CrossRef] [PubMed]
42. Petković, A.B.; Matić, S.M.; Stamatović, N.V.; Vojvodić, D.V.; Todorović, T.M.; Lazić, Z.R.; Kozomara, R.J. Proinflammatory cytokines (IL-1β and TNF-α) and chemokines (IL-8 and MIP-1α) as markers of peri-implant tissue condition. *Int. J. Oral Maxillofac. Surg.* **2010**, *39*, 478–485. [CrossRef] [PubMed]
43. Duarte, P.M.; Serrão, C.R.; Miranda, T.S.; Zanatta, L.C.S.; Bastos, M.F.; Faveri, M.; Figueiredo, L.C.; Feres, M. Could cytokine levels in the peri-implant crevicular fluid be used to distinguish between healthy implants and implants with peri-implantitis? A systematic review. *J. Periodontal Res.* **2016**, *51*, 689–698. [CrossRef] [PubMed]
44. Ghanem, A.; Abduljabbar, T.; Akram, Z.; Vohra, F.; Kellesarian, S.; Javed, F. A systematic review and meta-analysis of pre-clinical studies assessing the effect of nicotine on osseointegration. *Int. J. Oral Maxillofac. Surg.* **2017**, *46*, 496–502. [CrossRef] [PubMed]
45. Akram, Z.; Vohra, F.; Bukhari, I.A.; Sheikh, S.A.; Javed, F. Clinical and radiographic peri-implant parameters and proinflammatory cytokine levels among cigarette smokers, smokeless tobacco users, and nontobacco users. *Clin. Implant Dent. Relat. Res.* **2018**, *20*, 76–81. [CrossRef] [PubMed]

46. Khan, J.; Hassun, H.; Zusman, T.; Korczeniewska, O.; Eliav, E. Interleukin-8 levels in rat models of nerve damage and neuropathic pain. *Neurosci. Lett.* **2017**, *657*, 106–112. [CrossRef]Clin. Implant Dent. Relat. Res.
47. Khan, J.; Ramadan, K.; Korczeniewska, O.; Anwer, M.M.; Benoliel, R.; Eliav, E. Interleukin-10 levels in rat models of nerve damage and neuropathic pain. *Neurosci. Lett.* **2015**, *59*, 99–106. [CrossRef]
48. Buhlin, K.; Gustafsson, A.; Andersson, K.; Håkansson, J.; Klinge, B. Validity and limitations of self-reported periodontal health. *Community Dent. Oral Epidemiol.* **2002**, *30*, 431–437. [CrossRef]
49. Monje, A.; Pons, R.; Insua, A.; Nart, J.; Wang, H.L.; Schwarz, F. Morphology and severity of peri-implantitis bone defects. *Clin. Implant Dent. Relat. Res.* **2019**, *21*, 635–643. [CrossRef] [PubMed]

MDPI
St. Alban-Anlage 66
4052 Basel
Switzerland
Tel. +41 61 683 77 34
Fax +41 61 302 89 18
www.mdpi.com

Applied Sciences Editorial Office
E-mail: applsci@mdpi.com
www.mdpi.com/journal/applsci